Kouadio Yao Guerschom N'Gotta

Urban Dynamics and Land Management

Kouadio Yao Guerschom N'Gotta

Urban Dynamics and Land Management

In the villages integrated into the city: the case of SAPIA, TAGOURA and ZAKOUA in DALOA (Côte D'Ivoire)

ScienciaScripts

Imprint

Any brand names and product names mentioned in this book are subject to trademark, brand or patent protection and are trademarks or registered trademarks of their respective holders. The use of brand names, product names, common names, trade names, product descriptions etc. even without a particular marking in this work is in no way to be construed to mean that such names may be regarded as unrestricted in respect of trademark and brand protection legislation and could thus be used by anyone.

Cover image: www.ingimage.com

This book is a translation from the original published under ISBN 978-613-8-43610-2.

Publisher:
Sciencia Scripts
is a trademark of
Dodo Books Indian Ocean Ltd. and OmniScriptum S.R.L publishing group

120 High Road, East Finchley, London, N2 9ED, United Kingdom
Str. Armeneasca 28/1, office 1, Chisinau MD-2012, Republic of Moldova, Europe
Printed at: see last page
ISBN: 978-620-5-80905-1

Contents

FOREWORD

As part of our university training, we are submitted a тёгоие of research in Master 2, in order to obtain the MASTER degree.

The theme of our dissertation is: "*Urban dynamics and land management in villages integrated into the city: the case of Sapia, Tagoura and Zakoua in Daloa (Cote d'Ivoire)*".

Far from being an exhaustive study, our work is a modest contribution to the numerous researches already made on urban land tenure in Ivory Coast. This study focuses on all Ivorian cities, although the in-depth study is based on the example of the city of Daloa.

ACKNOWLEDGEMENTS

We would like to express our sincere thanks to all those who contributed to our training, to the realisation of our research work or to the production of this thesis.

We thank the President of the Jean Lorougnon Guede University, Professor TIDOU Abiba Sanogo epse KONE, the Director of the Training and Research Unit of Social and Human Sciences, Professor KONE Issiaka, the Vice-Director, Dr MAFOU Kouassi Combo and the Head of the Department of Geography, Dr OUATTARA Sahoti for the quality of the supervision.

We sincerely thank Professor KOFFI Brou Emile, our Scientific Director, and Dr YAO Kouassi Ernest, our supervisor, for their availability and their interest in our subject. We also thank Dr TRAORE Kinakpefan Michel and Dr GUELE Gue Pierre for their observations which were very useful.

We are grateful to all the professors of the Geography Department and those of the History Department to whom we address our thanks for the quality of the training we have received. We also thank Dr TRA Bi Crolaud Sylvain of the UFR Agroforestry for his encouragement.

Nous associons a ces remerciements M. GRAGBO Pepe Felix, M. GROGUHE Lago et M. HANON Digbeu, respectivement chefs de Zakoua, Tagoura et Sapia, M. HONE Jacques, Secretaire General de la notabilite de Sapia ; M. YAO N'Goran Bazin, Directeur General du Cabinet International d'Appui au Developpement Local ; M. ASSAMOI Diby Albert, President of the Daloa Commune Landowners' Association; Mr. KOUAKOU Severin of the Topography and Subdivision Department of the Regional Directorate of the Ministry of Construction and Urbanism; Mr. YAO Kouame Simplice of the Cadastral and Tax Department; for the quality of the information provided to us.

We sincerely thank Mr. AKOTO Assamoi Ernest, Secretary General of the APTDC, and Director General of the Agence Immobiliere et Services en Cote d'Ivoire, Mrs. YAPI Thierry Anne Delphine epouse AKOTO ; Mr. GBALOU Joachin ; Mrs. AKPOUE Amenan Christine ; for all their support

We thank Mr. N'GOTTA Kouadio Remi, our father, and Mrs. KORO Amenan Rebecca, our loving mother, and the entire N'GOTTA family whose love and support is more than vital.

Finally, we would like to thank all those who, from near or far, contributed to the realisation of this work, whose names are not mentioned here.

SUMMARY

With its remarkable spatial extension, the city of Daloa has now reached the periurban villages of Sapia, Tagoura and Zakoua. It is a question of the customary landowners making their plots available. The present study aims at knowing the impact of the urban dynamics of Daloa on land management in the integrated villages of Sapia, Tagoura and Zakoua. The methodology adopted is based on the exploitation of an abundant literature on the subject, complemented by field visits, interviews with city managers (Regional Director of Construction and Urbanism, Mayor) and a questionnaire survey of landowners in Sapia, Tagoura and Zakoua. The new subdivisions are opening up on plots of land that have already been the subject of transactions, either between indigenous Bëtë people and people from outside the region (allochthones and allogenes), or between the Mairie and the applicants for building land. The subdivision operations and the sale of land, supported by strong land speculation, lead to land disputes that weaken social cohesion. As a result of these disputes, the plots are insufficiently developed, thus making the urban landscape ugly.

Keywords: Daloa, urban dynamics, integrative logics, periurban villages.

1. UNDERSTANDING OF THE SUBJECT

The process of change and transformation of the modes of use of space, modes of economic production and modes of consumption, is reflected in the Ivorian territory in two ways: the demographic expansion and the multiplication of cities. Although declining, the annual growth rate of the urban population remains high: from 3.8% over the 1975-1988 period, the annual growth rate rose to 3.3% between 1988 and 1998, then to 2.6% between 1998 and 2014 (RGP 1975; RGPH 1988, 1998 and 2014). Rural settlements quickly outgrow their large villages and become towns, either through infrastructure development, or by the elevation of the administrative level of territorial organisation or by the intensification of trade activities. These cities expand on their margins and reach the peripheral villages. The integration of the peripheral villages into the city requires the customary owners to make their portions of land available.

Moreover, highly centralised since the colonial period, urban land management in Côte d'Ivoire has undergone a significant evolution since 2013. Until 2013, administrative subdivisions were initiated by sub-prefectures and town halls: the town hall was in charge of subdivisions within the communal perimeter while the sub-prefecture intervened outside it. A commission for the allocation of plots was set up, comprising representatives of the prefecture, the sub-prefecture or the town hall, depending on the case, the land registry and the building department. In 2014, the state consolidated customary rights and gave the possibility to village communities and holders of customary land rights to initiate the subdivision of their plots. The effect of this new regulatory provision was to increase the power and role of indigenous people, putting them at the heart of land production.

In Daloa, the new housing estates open onto plots belonging to customary owners in the peri-urban villages of Sapia, Tagoura and Zakoua. However, 80% of these plots are used for agricultural purposes or have been ceded to non-natives and/or allogenes who use them for this purpose (Our survey, 2018). Thus, the land heritage of these integrated villages is subject to strong pressures due to the competition between the actors in land production. This study aims to show the impact of the urban dynamics of Daloa on land management in the integrated villages of Sapia, Tagoura and Zakoua.

2. JUSTIFICATION OF THE CHOICE OF THE SUBJECT

Three reasons justify the choice of our subject entitled: "Urban dynamics and land management in villages integrated into the city: the case of Sapia, Tagoura and Zakoua in Daloa (Cote d'Ivoire)".

The first reason is scientific. Indeed, the study focuses on the phenomenon of urbanisation and its implications on the surrounding rural area. This thesis is about the impact of the integration of peri-urban villages into the city of Daloa. In addition, the University institution has the vocation to promote local and universal development. Hence the interest given to the problems of development encountered by the city of Daloa, which houses the Jean Lorougnon Guede University.

The second reason is social. At the social level, the problems posed by the integration of villages into urban areas are considerable and perceptible at all levels. The integration of the villages of Sapia, Tagoura and Zakoua into the city of Daloa poses problems of land use and management. Also, the populations are exposed to problems of access to land, housing, equipment and infrastructure and to problems of access to a healthy living environment.

As scientific research is a decision-making tool for territorial administrators, this thesis could serve as a guide for decision-makers and managers of Ivorian cities in general, and of the city of Daloa in particular. In addition, this study could constitute a database for the development and better management of Ivorian cities.

The third reason is personal. Indeed, it is the role of the researcher to theorise everything that appears as empirical data that can be observed. Consequently, the effects of the changes in the rural space are perceptible and irreversible. These changes have an impact on the socio-economic and spatial-environmental development of both the changing rural areas and the growing city. All these things have animated our curiosity and led to the formulation of this research topic.

3. REVIEW OF LITERATURE

Worldwide, and more particularly in developing countries, the relationship between urban sprawl and land management is a major development concern; the land issue being at the heart of all territorial development processes. In order to understand this study, we have reviewed the literature on the following themes: factors of urban dynamics; urban sprawl; land management; transformation of peri-urban spaces.

3.1. Factors of urban dynamics

Urban dynamics is a phenomenon closely linked to demographic growth. Thus, HAERINGER P. (1977) states that: "Abidjan's expansion was rapid because it had to cope with several population doublings without too much difficulty". Similarly, CHALINE (2000, p. 21) explains that the transformations of cities depend both on exogenous factors, the importance of which is well illustrated by migratory flows, and on an internal dynamic which is expressed in particular by residential mobility and by progressive changes in the social occupation of intra-urban space.

According to VERON J. (2008, p.14), 'the city grows by the arrival of migrants wishing to escape from rural poverty, the demographic weight of the cities increases, the decision-makers who live there tend to favour the cities, when allocating public resources, the cities become even more attractive, new migrants arrive in the city, the political power of the cities is strengthened. Thus, DURREAU F. (2008) explains that "the proportion of the population living in cities will increase in the coming decades, and even if migration from the countryside to the cities decreases, the youthful age structures of the populations already living in cities will maintain high rates of growth, higher than those observed in rural areas".

As for VERON J. (2008, p. 4), he indicates that "a country in which the proportion of urban dwellers is low is also a country that is not very advanced in its transition to a higher fertility rate. However, the relationship between total population growth and the rate of urbanisation could be the opposite: the lower the degree of urbanisation of a country, the faster the population growth, and the less advanced the demographic transition in a less urbanised country'.

This view is corroborated by Miguel Villa (1996) who distinguished Latin American countries according to their levels of urbanisation to see what link could be established with the degree of progress of the demographic transition. The most urbanised countries, such as Argentina, Chile and Uruguay, with rates above 85% at the end of the 1980s, were the most advanced in the demographic transition. Less urbanised countries such as Guatemala, Haiti and Honduras (rates below 45%) were then only beginning their demographic transition. Countries with intermediate levels of urbanisation showed some diversity, as their degree of progress in the demographic transition did not correspond to the blërarcble of urbanisation rates.

According to the European Environment Agency (2006 report, p. 6),   "the Urban sprawl occurs in a given area when the rate of land use and consumption for urbanisation is faster than population growth over a given period. It is caused more by changes in lifestyles and consumption than by population growth.

Furthermore, urban dynamics are affected by the development of economic activities. This justifies the words of ATTA K. (1978) who indicates that the city results from the inscription on the field of economic growth: "urban spaces have always been organised around one or two major roads that cross the city in width and length". Continuing, he states that 'the spatial extension of the city of Bouake was made from the railway station, which took over the role of the city's centre of gravity from the military post. This spatial extension is also the result of the strong attraction of the city linked to its role as a regional administrative centre and of a strong immigration (ATTA K. 1984).

According to KADET B. (1999), the dynamics of the extension of Guiglo was possible thanks to the economic and social development policy of the Ivorian state. Guiglo benefited from important investments at the end of the 1960s (FRAR, FIAU and five-year plans). In this regard, FUJITA and THISSE (2003) indicate that 'agglomëration would be the territorial counterpart of economic growth'.

As for BEAUCIRE F. et al. (2007, p. 22), they argue that "if in the context of the development of service activities, jobs tend to be polarised in the centre of the agglomëration, the subsëquent land costs tend to relegate the space-consuming activities to the përiurban area".

On the other hand, POLESE M. (2010, p.5), explains that "the positive relationship between urban size and productivity is only true within a country, not between countries. The positive effect of urban size disappears as soon as national differences in development are taken into account. The contribution of agglomëration (by way of urbanisation or urban growth) to increases in productivity is static through the spatial distribution of resources." Similarly, VERON J. (2008, p. 2) argues that

"The accumulation of population in ever larger cities would have become a largely autonomous, even "antiëconomic" phëne, especially in developing countries".

3.2. Urban sprawl

Urban sprawl is a geographical phenomenon that has interested many authors. According to Philippe JULIEN, (2005, p. 3), "Urban sprawl" is commonly understood as the result of the propensity of inhabitants to settle, preferably in suburban housing, at the periphery of cities. This is sometimes the result of individual choices, such as the desire to have a garden (usually small in the case of housing estates) or to avoid the inevitable promiscuity of urban collective housing. It can also be the result of constraints: for example, the high cost of land in the city centre sometimes makes it difficult for a growing family to access decent housing.

Similarly, CUNHA A. et al (2000, p. 16) explain that urban sprawl is a process of dispersion of the built environment and expansion of the urban space, successively embracing rural "hinterlands" through annexation and incorporation. It is marked by three generally correlated developments: the growth of the footprint of urban components (buildings, civil engineering structures, transport infrastructure, etc.); the dispersion of urban elements in the territory; the multiplication, enlargement and lengthening of urban transport networks.

This opinion is also shared by ENAULT (2003) who indicates that just like a classic diffusion process, urban sprawl therefore proceeds by contagion. The transformation takes place by simple contact between the agglomeration and the countryside but also by 'heating'. Thus, the urban centre polarises a large rural area whose 'growth potential' decreases the further away one gets from the city centre. Like a forest fire, the city consumes the nearest areas while sending flaming mëches over long distances. The latter cause new secondary fires acting as the main focus.

According to GEDDES (1949), "every city tends to grow: the very function for which the city was created takes on an increasing importance; other functions are added to it. Each function requires more and more personnel; immigration leads to an inflation of functions and personnel. Then the rise in the standard of living requires more and larger accommodation". Continuing, he explains that this increase is done in various ways, either by crowding in or by projecting outwards. This depends on the ëpoons, technical power and йпапйёге, but also on the relations between the city and its hinterland.

BEAUJEU-GARNIER (2006) states that it is a 'projection of the city out of itself. It is a process in constant evolution'. Thus, 'urban sprawl can be described as the densification of territories located further and further from the heart of the city' (Philippe JULIEN, 2005, p. 3). On their side, BEAUCIRE F. et al. (2007, p.7) think that "I'ctalement urbain caractere le phCnomëne de croissance de l'espace urbanise de fagon peu maitrisce, produisant un tissu urbain tres lache, de plus en plus cloignc du centre de l'aire urbaine dont il est dëpendant".

For DERRUAU (1998), urban dynamics has three aspects: a spatial aspect, a functional aspect and a demographic aspect. The spatial aspect is concerned with the stages of extension of the city. The functional aspect concerns the succession of functions of adaptation to the historical circumstances which have led to the present city. The demographic aspect is the settlement of a population due to the functional aspect of the city.

In his study on the impact of industrial wood processing units on urban development in Daloa, YAO K. E. (2014) explains that the spatial dynamics of the city of Daloa are remarkable. The scarcity of land forces applicants in a hurry to build a roof over their heads to settle temporarily on the margins of the city in the hope that a possible subdivision would make them definitive owners. In doing so, these temporary occupants participate in the spread of the city to its margins.

Going in the same direction, KOUAME G. et al. (2016, p. 24) explain that "in the District of Abidjan, accelerated urbanisation is taking place at the expense of many villages. This phenomenon generally leads to the displacement of the local populations in the areas concerned. The subdivisions, which are the result, give rise to financial compensation intended to ensure the resettlement of the populations concerned by such operations.

In our study, the concept of urban sprawl is synonymous with urban growth, the extension of the city.

3.3. The transformation of periurban spaces

According to Pierre GEORGE and Fernand VERGER's dictionary of дёодгарЫе (2004, p. 314), the periurban is 1 sitiic rural space in përiphërie of a city and its suburbs and which is the object of deep landscapeëres, functional, dëmographic, social, cultural, and even political transformations.

In our work, the villages of Sapia, Tagoura and Zakoua are incorporated into the city of Daloa, hence the use of the concept of integrated village or village-district.

According to CUNHA A. (2000, p. 17),   "the përiurbanisation is characterised by the emergence of discontinuous urban configurations of the morphological agglomeration, close to the përiphërie of the suburban cores and characterised by low density, low diversity (sociodëmographic), but also by good accessibility to the rest of the urban space Përiurbanisation is a phënomëne ëtroitement Hë to transport development. For DESJARDINS X. (2017, p. 4), "periurban development is the most visible aspect of a transformation of all urban spaces under the effect of speed that has become accessible to the greatest number".

According to SERRANO J. (2007, p. 3), "the growth of urban centres tends to be transferred to the peripheral spaces rather than to the central spaces. This results in particular spaces, as much in their morphology (the habitat is not very dense), as in their function or their relationship to the city (their easy accessibility allows daily migrations but mainly by car), which are the përiurban spaces". This thought corroborates that of DAVODEAU H. (2005, p. 3) who states that "the process of përiurbanisation is characterised by urban diffusion in successive rings and is based on two main mechanisms: land imirclK' and daily work motives".

For JULIEN P. (2005, p. 3), "periurbanisation is the result of the spatial extension of housing but also of transport infrastructures and activities: these extensions should also be considered in parallel with that of housing".

As for PRAT A. (2010, p. 22), he affirms that 'the urban fringe is coveted by all city dwellers, those who are in search of lëgalised land, and others, residing in regular allotment areas and who wish to accumulate several plots within their family group, thus building up a land heritage that testifies to their urban integration'.

Furthermore, përiurbanisation is a phënomëne Hë to the development of new lifestyles. For EGGERICKX et al (2002), 'l espace përшгЬат appears to be socio-ckmograplitically privilëgië, a direct consëquence of the flambëes of the land and real estate m;irclers that caracerise it, but within which strong disparities of income appear'.

DESJARDINS X. (2017, p. 4), ënounces that "the гёаШё рёпигЬате: it is not only the extension of the "urban spot" or the increase of suburban areas, it is also 1 extension of "urban living areas" that agregate larger and larger rural areas".

For AYDALOT (1985), përiurbanisation is Пёс' to the development of new lifestyles, especially to the development of the single-family home and often involves high rates of private motorisation ёs.

For VANIER M. (2000, p. 106), 'the third space [...] refers to a hybrid, complex space, shared between cities and the countryside, belonging to logics that are both urban and rural, in other words neither strictly one nor the other. These inhabitant, social, economic, technical logics, etc. cannot be taken into account exclusively in the context of the urban environment. These inhabitant, social, economic, technical, etc. logics cannot be taken into account exclusively in the name of the city or in the name of the countryside, and by their respective system of action: where the urban society purposely mixes the urban and rural components of its flourishing, it is very likely that collective action, and within it public action, must find their effectiveness in the same mix, in other words, in the Tinter-territorial^.

According to DAVODEAU H. (2005, p. 3), "contemporary urban landscapes are presented as a fluctuating and porous glaze, an assembly of two worlds that exist in our representations only by their opposition to each other. Fragmented, incoherent, fragmented, temporary, juxtaposed, fluctuating, discontinuous, intermëdiary, they dëstabilize our reading grid of landscapes (rural and urban) by making another гелШё emerge, composed of both agrarian (roads, plots and farms, hamlets) and urban (housing estates, activity zones, infrastructures and facilities) landscapes".

DESJARDINS X. (2017, p. 3) points out that while dens^ and divers^ in proximo are the hallmarks of the historic city, new urban^s are being invented thanks to the automobile and new information and communication technologies in përiurban spaces.

According to JAUZE J-M. and NINON J. (1999, p. 4) periurban areas are the scene of a high demographic growth based on a positive migratory balance. DUPONT V. (2005) states that 'understanding peri-urban dynamics is an essential prerequisite for any urban and regional planning undertaking'.

3.4. Land management

ATTA K. (1984) argues that Cote d'Ivoire's land legislation makes the state the sole owner of the land. It is the state that by l'intermëdiary of the subdivision procedures changes the rural soil into urban soil.

Taking the example of Abidjan, YAPI-DIAHOU (1981) points out that Ebrie land management was ensured by three levels of power and authority which were the 'Nana', the village chief and the head of the family. The introduction of the notion of 'vacant and masterless land' by the settlers was to encourage the confiscation of land from the indigenous populations for the benefit of the rulers. Continuing, YAPI-DIAHOU (1990) explains that the creation of the Societe d'Equipement des Terrains Urbain (SETU) was a response to the need to subdivide, develop and improve the urban space in order to allow SICOGI and SOGEFIHA to build several housing units in Abidjan.

According to KOUAME G. et al (2016, p. 25), 'in Abidjan, there is the coexistence of two organisations, traditional and modern, namely the customary authority(ies) and the administration, which have distinct procedures for land governance. The result is an overlapping of the two jurisdictions and a prejudicial concurrent management with a real risk of disputes.

As for AKPINFA E. (2006, p. 48), he points out that in Benin, subdivision appears much more often as an operation to regularise the anarchic occupation by the populations of the urban peripheries, an operation to reorganise and restructure the land in urban and periurban areas. He went on to explain that the neighbourhood chiefs and traditional or customary chiefs are intermediaries between the administrative authorities and the households. In land matters, they are important sources of information for the validity of a property. In the case of land sales between individuals, they are the first level of officialisation of the sale procedure.

ZIAVOULA R., (1987, p. 2) states that in Brazzaville, the non-existence of a public land registry allows landowners to choose the purchaser, often according to the degree of relationship that exists between them and the person presenting the potential purchaser.

The person who serves as intermediary constitutes a moral guarantor for the landowner.

UN-HABIAT (2011) states that "governments should strengthen their capacity to respond to the pressures of rapid urbanisation. Particular attention should be paid to land management in order to ensure its economic ë use, protect fragile ëcosystëmes and facilitate the poor Гассёз to land in both urban and rural areas".

Furthermore, the price of urban land is a central element of land management. On the subject of land value, WINGO's (1962) model highlights the importance of transport. "The worker who lives closer to the city бёпёйас a situational rent, his location appears more favourable and the land value of the place where he resides will be more ёкуё than that of a more distant location. As the size of the city increases, so does the land value and residential density. An improvement of transport reduces the inequalities and more or less eliminates the denounced consequences.

For ALONSO (1965), "the improvement of transport generally favours

This would lead to a përiphë of the agglomeration and a rise in land prices in this area, while land pressure would decrease in the centre.

According to MAYER R. (1965), 'the price of land on the periphery is equivalent to the price of agricultural land plus the costs of servicing and equipment, plus a sum corresponding to an anticipation of the value acquired by the land that can be urbanised, plus a rent when the building land is large'. Thus, BEAUCIRE F. et al. (2007, p. 21) point out that 'the opening up of new areas to urbanisation represents a financial windfall for owners of agricultural land, when the farms are in difficulty or no longer have any buyers'.

NGANA F. et al (2010, p. 2) argue that 'real estate speculation intensifies with the expansion and growth of the urban population'.

In addition, land conflicts arise in land management. ZIAVOULA (2007, p. 3) explains that the sale of land to several people is a major cause of land disputes. The landowner always hopes to get another buyer to make a better deal. Thus, the chain of conflicts between the landowner and the different acquirers is established. Even in situations where the landowner manages to pay back his previous clients, the conflict remains between the various acquirers.

Taking Korhogo as an example, KOUAKOU B. et al. (2015, p. 8) explain that 'the most frequent cases of land disputes concern members of the same family opposed by the succession system, notably the nephews and sons of the landowner'. Thus, conflicts within villages are rooted in the sharing of family land (Atlas of Population and Equipment, 2007).

Land tensions make it difficult to develop the acquired portions of land. Jean-Philippe Plateau (1998) explains that 'the tensions surrounding land, observed at the periphery of cities, reflect the anxiety of local actors to see their land reserves disappear'.

At the end of the analysis of the literature consulted, we note the insufficiency of data relating to land management in the villages integrated into the cities, particularly in Côte d'Ivoire after 2014. However, due to its remarkable spatial extension, the city of Daloa has integrated the villages of Sapia, Tagoura and Zakoua. In addition to the customary landowners, intermediaries who improvise themselves as developers or entrepreneurs are involved in land production in these integrated villages. The pressure on customary plots of land leads, in some places, to land disputes. Hence the interest of the present study.

4. PROBLEMATICS

The population growth and the ever-increasing number of people seeking land in Daloa are increasing the pressure on building space. Today, the city of Daloa has reached the peri-urban villages of Sapia, Tagoura and Zakoua. Moreover, 1 Etat a travers i'arrete n°20/MCLAU/DGUF/DDU/lgb du 20 Fëvrier 2014, donne la possibilite aux communautës villageoises et aux dëtenteurs de droits coutumiers fonciers dë initier le lotissement de leurs parcelles. In Daloa, the new subdivisions open up on plots that have already been the subject of transactions, either between indigenous Bëtë and people from outside the region (allochthones and allogënes), or between the Mairie and the applicants for building plots. The subdivision operations and the strong land spëculation that results from them lead to land disputes that weaken social cohesion. Also, due to land disputes, plots are insufficiently developed, thus marring the urban landscape with scrub and wasteland, sometimes used for intra-urban agriculture. The question is: what is the impact of urban dynamics on land management in the integrated villages of Sapia, Tagoura and Zakoua?

From this central question flow the following subsidiary questions: How do the spatial dynamics of the city of Daloa operate?

How is land managed in 'integrated villages'?

What is the socio-economic and spatio-environmental impact of the urban dynamics of Daloa on land management in the integrated villages?

5. OBJECTIVES OF THE STUDY

5.1. General objective

This study aims to understand the impact of urban dynamics on land management in integrated villages.

5.2. Specific objectives

Specifically, this study aims to :

- Describe the dynamism of the urban space of Daloa;

- Studying land management in the "integrated villages" of Sapia, Tagoura and Zakoua;

- To analyse the socio-economic and spatio-environmental impact of the urban dynamics of Daloa on land management in the integrated villages.

6. RESEARCH METHODOLOGY

The methodology adopted can be summarised in distinct stages: formulation of the research hypotheses, identification of the variables to be analysed, presentation of the geographical framework of the study, data collection (documentary research, direct observation in the field, interviews, questionnaire survey), data processing and analysis.

6.1. Research hypotheses

6.1.1. General hypothesis

The urban dynamics of Daloa has a socio-economic and spatial-environmental impact on the management of the inegre villages SAPIA, TAGOURA and ZAKOUA.

6.1.2. Specific assumptions

6.1.3. Specific assumption 1:

In view of the remarkable spatial dynamics of Daloa, the villages Sapia, Tagoura and Zakoua are now integrated into the city.

6.1.4. Specific assumption 2:

Land production in the integrated villages of Sapia, Tagoura and Zakoua is carried out according to an internal organisation and land division operations.

6.1.5. Specific assumption 3:

Land speculation is at the root of land disputes and the rise of undeveloped plots that are marring the urban landscape of Daloa.

6.2. Analysis variables

To verify our hypotheses, we used qualitative and quantitative variables related to the following themes: demographic dynamics, spatial dynamics, land production, and the purchase and sale of peri-urban land.

Table 1: Variables related to population dynamics

Qualitative variables	Quantitative variables
-Gender - Nationality - Marital status - Socio-professional categories	- The volume of population - Population density - The average annual growth rate of the population

These variables allowed us to describe the evolution of the population of Daloa in order to assess its impact on the expansion of the city.

Table 2: variables related to spatial dynamics

Qualitative variables	Quantitative variables
- The Daloa master plan - The extension sectors of the city - The land use plan - Land use - The pace of housing construction - Neighbourhood typology - Housing typology - The nature of the building materials - The mëcanisms of spatial extension	- Number of housing estates built - Number of approved allotments - Number of lots developed - Number of districts - Area of urbanised space - Land use rate

These variables allowed us to know the areas and mechanisms of extension and the city of Daloa.

Table 3: variables related to land production

Qualitative variables	Quantitative variables
- Socio-professional profile of land production actors	- Duration of developments - Cost of subdivisions - Number of housing estates built

- Role of the actors - Typology of housing estates - The subdivision process - The legal provisions in force	- Number of lots per development - Size of the parcels developed - Number of infrastructures and facilities

These variables allowed us to know the mëcanisms of land production and to appreciate the Гопаёre pressure in the integrated villages Sapia, Tagoura and Zakoua.

Table 4: variables related to the purchase and sale of periurban land

Qualitative variables	Quantitative variables
- Identification of field vendors - Conditions of access to land - How to buy - Documents to be provided for the purchase - Documents obtained at the time of purchase - Role of the Ministry of Construction - Difficulties encountered in obtaining a lot - Procedure for obtaining the CDA - Types of land disputes - Method of resolving land disputes - Geocultural origin of the buyers - Function of the acquirers - Level of education of the purchasers	- Date of purchase of the land - Lot size - Land prices - Timeframe for the purchase process - Number of lots sold - Number of lots purchased - Age of the purchasers - Time limit for the development of the lots

These variables enabled us to understand the socio-professional profile of land sellers and buyers, the procedures and conditions for buying or selling land, and the difficulties involved.

6.3. Geographical framework of the study

Located in the centre-west of Ivory Coast, the city of Daloa is the capital of the Haut-Sassandra region. It is 383 km from Abidjan, the economic capital of the country, and 141 km from Yamoussoukro, the political capital. Located at 6°53' north latitude and 6°27' west longitude, Daloa has an area of 5305 km^2 with a population of 245,360 inhabitants (RGPH, 2014). Daloa's geographical location at the crossroads of national and international roads and the combination of university, military, administrative and commercial functions, as well as its humid climate, make the city attractive. Figure 1 shows the location of the city of Daloa in the Haut-Sassandra region.

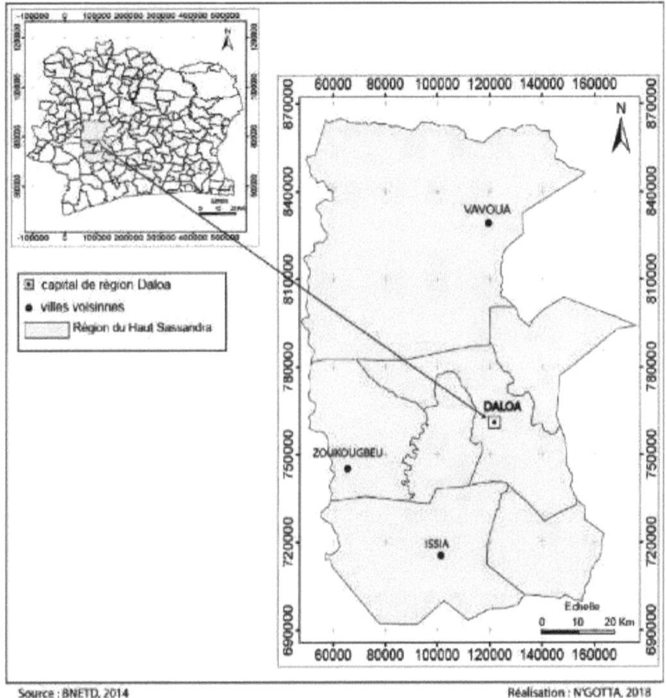

Source : BNETD, 2014   Réalisation : N'GOTTA, 2018

Figure 1: Location of the city of Daloa in the Haut-Sassandra region

6.4. Rationale for the choice of survey sites

The choice of Sapia, Tagoura and Zakoua localisës respectively at the eastern, northern and southern përiphëries of Daloa (Figure 2) allows us to appreciate the changes made in urban land management. Indeed, Sapia, Tagoura and Zakoua have been subject to administrative allotments тШёз by the town hall: Sapia in 1984 and 2012; Tagoura in 1974; Zakoua in 1982 with an extension in 2007. Since 2014, these villages have been experiencing private housing developments at the initiative of landowners. It is also a question of village communities and customary land rights holders making their land holdings available to the state. Due to the high demand for building land, customary landowners initiate pricey subdivision operations and make the lots available to the population, who buy land to build houses on. However, land disputes are recurrent in these villages.

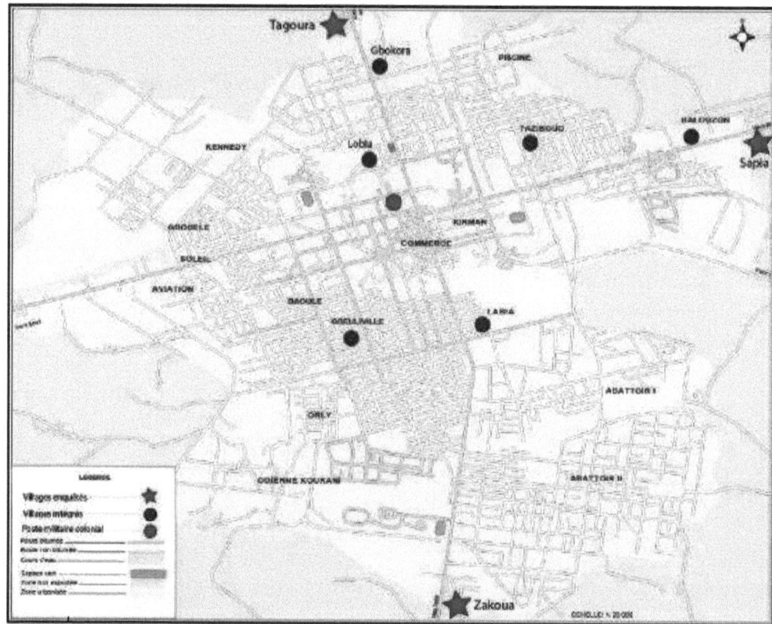

Source: ¥A0 2014; NDTL, 2015   Realisation:N'GOTTA,2019

Figure 2: Presentation of the survey sites

6.5. Scales of observation

The city of Daloa and the integrated villages Sapia, Tagoura and Zakoua constitute our observation scales.

6.5.1. The city of Daloa

This scale of observation allows us to highlight the spatial dynamics of the city of Daloa and the mechanisms of space consumption.

6.5.2. Integrated villages

At this scale, we studied the production of land and the impact of urban sprawl on land management. This involved observing and assessing the size of lots, built and unbuilt lots, disputed land, the procedure for buying and selling land, etc.

6.6. Data collection techniques

Data collection was carried out through desk research, direct observation in the field, interviews with resource persons and a questionnaire survey of landowners.

6.6.1. Literature search

We visited the libraries of the University Jean Lorougnon Guede, the municipal library of Daloa, the Institute of Tropical Geography (IGT) and the Institute of Research for Development (IRD), in order to obtain the documentation necessary for our study; not forgetting the Internet sites that we consulted. In addition to methodological works, we consulted books, theses, dissertations, journals, reports and articles on urban dynamics and land management.

The works of GUMUCHIAN H., MAROIS C. and FEVRE V. (2001), *"Initiation a la recherche en geographie: Amenagement, developpement territorial, environnement"*, SCHEIBLING J. (2004), *"Qu'est-ce que la geographie?"*, BAILLY A. and FERRAS R. (2004), *"Elements d'epistemologie de la geographie"* have been useful to us in defining the steps of the research and the methodology for writing the dissertation

The works of FOFANA B. (2015), *"Dynamiique urbaine et problemes environnementaux a Bouake"*, COTTEN A-M. (1974), *"Un aspect de l'urbanisation en Cote-d'Ivoire"*, DJEKI J. (2003), *"politique urbaine et dynamique spatiale au Gabon : le cas de Port - Gentil"*, LASSERRE G. (1970), *"La dynamique de l'espace urbain a Libreville : regiementation fonciere et morphologie des quartiers"*, SAID B. (2008), *"La ville en question - analyse des dynamiques urbaines en Algerie"*, ALLA A. (1991), *"Dynamique de l'espace periurbain de Daloa"* explain to us the urban dynamics opërës in the cities of Bонaкë and Daloa (Cote d'Ivoire), Port Gentil (Gabon), Libreville (Congo) and the cities of Algeria.

The scientific productions of KOFFI E. B (2010), *'Les lotissements irregularities and the production of the city: the Ayakro and Sagbe districts in Abidjan"*, MEMEL F. A (2006), *"Dynamisme urbain et gestion fonciere a Grand Bassam"*, YAON B. (2009), *"La Maitrise du foncier dans le Processus du Programme spécial de Transfert de la Capitale a Yamoussoukro"*, d' AKPINFA D. E (2006), *"Problematique de la gestion fonciere dans les centres urbains secondaires du Benin"*, YOUANA J. (1996), *"Gestion fonciere et discipline urbanistique au Cameroun : Apports et limites du permis de construire"*, explain to us how urban land tenure is дёгë in Abidjan, Grand Bassam, Yamoussoukro (Cote d'Ivoire), Benin and Cameroon.

6.6.2. Direct observation in the field

We visited the housing fronts of the integrated villages Sapia, Tagoura and Zakoua. Direct observations in the field allowed us to appreciate the dynamics of the extension of the city of Daloa and the spatial and environmental impact of the extension of the city on its margins.

Our attention was focused on the areas of town extension and the rate of land occupation. We also observed the dimensions of the lots, the unbuilt lots, 2'ëlal of subdivision servicing, the infrastructures built on the administrative reserves, the existence and functioning of the roads and miscellaneous networks (V.R.D).

During the visits, photographs were taken and analogue data such as graphs and maps were produced as evidence of our observation of the field.

6.6.3. Interviews

Interviews were held with administrative officials, including the Mayor and the Regional Director of Construction and Urbanism; a geomëtre expert agreeing to Daloa, the traditional chieftainships of Zakoua, Tagoura and Sapia, the Director General of Amënagement Immobilier et Service en Cote d'Ivoire (AISCI) and the Director General of Cabinet International d'Appui au Dëveloppement Local (CIADEL), on the other hand.

The choice of administrative officials (Mayor and Rëgional Director of Construction and Urbanism) is justified by the fact that they are the managers of the town. The discussions focused on the logic of integrating the villages рёйигЬятз into the city, in particular the procedures for subdividing and purchasing land, the provision of infrastructure in the "village-districts", the types of land disputes and their methods of resolution, the actors involved in the conflict, and the social, economic and spatial-environmental impact of these land disputes on the development of the city.

Interviews with the chiefdoms of Zakoua, Tagoura and Sapia were necessary in order to understand the socio-economic impacts of the integration of the villages рёгшАятз into the town. The discussions focused on the role of the chieftaincy and land management committees in the subdivision operations, the way the lots are shared, the size and price of the plots, the economic activities, the payment of land taxes, the existing land disputes in these villages and their modes of resolution, the actors in conflict.

The interview with the Director General of Amënënagement Immobilier et Service en Cote d'Ivoire is explained by the fact that it is a prize structure that has financed housing development operations in the three villages (Zakoua, Tagoura and Sapia) as a project owner. The interviews focused on the legal and economic framework of the subdivision operations and the difficulties encountered, but also on the size of the lots and their selling prices, the types of land disputes and their resolution, the actors in conflict and the social, economic and spatial-environmental impacts of these land disputes on the development of the town.

The interview with the General Director of the Cabinet International d'Appui au Dëveloppement local focused on issues of urban governance in general, and on the management of the city's habitat and environment by the Mairie in particular.

6.6.4. Questionnaire survey of customary owners

We proceededëdë to a questionnaire survey among customary landowners in order to collect information on the management of their plots. Landowners are people who own a piece of land. Questions were asked about the delimitation of plots, the working agreement with the land surveyor or real estate agency, the division of plots and the use made of them, land disputes, the size and price of plots.

The size of the sample surveyed was determined by the reasoned choice method. This choice is justified by the fact that there is no statistical data on the number of landowners in the integrated villages. Indeed, not all village inhabitants are landowners. We therefore conducted a systematic enumeration of landowners in each village. In total, 1503 landowners were counted in the three villages, including 560 in Sapia, 583 in Tagoura and 360 in Zakoua (Table 5).

Table 5: Distribution of surveys by village

Villages	Total population of the village (RGPH, 2014)	Total number of landowners	Number of landowners interviewed (10% of total)
Sapia	1268	560	56
Tagoura	1405	583	58
Zakoua	1001	360	36
Total	**3674**	**1503**	**150**

Source: personal survey, 2019

We chose to interview 10% of the total number of registered landowners. The sample was 150 landowners, of which 56 in Sapia, 58 in Tagoura and 36 in Zakoua.

The criteria for selecting the surveys were as follows:

- The age of the landowner (at least 30 years);
- The profession (civil servant, private individual) ;
- The number of lots the landowner has (the landowner must have at least five lots).

In each village, we selected landowners by large family. Taking into account the occupations of the respondents (farmers, civil servants, shopkeepers, craftsmen), we defined three periods for applying the questionnaires. We chose the morning from 7am to 9am, the afternoon from 12pm to 2pm and from 5pm to 7pm. This method allowed us to reach the quota indicated for each village.

6.7. Data processing

The information collected was processed in an analytical and computerised manner. Quantitative data were statistically processed and the results presented in tables and figures using Excel and SPSS software.10 Qualitative data were analysed and used to explain and illustrate the work. The data is expressed in tables, figures, photos and maps. Adobe Illustrator 11.0.0 and Arcgis 4.4 were used to produce the maps.

The results of our research are based on three major axes:

The first axis describes the dynamism of the urban space of Daloa.

The second axis ëstudies land management in the 'integrated villages' Sapia, Tagoura and Zakoua.

The third axis analyses the socio-economic and spatio-environmental impact of the urban dynamics of Daloa on land management in the integrated villages.

Table 6: Summary table of the research

Findings	Questions from research	Objectives	Assumptions	Analysis variables	Method of data collection	Expression of results	Plan
The city of Daloa extends over its margins and reaches the villagesperiurbans Sapia, Tagoura and Zakoua.	How does the dynamism of urban space Daloa ?	Describing the dynamis mof the urban space of Daloa	With regard to the remarkable spatial extension of Daloa, the villages Sapia, Tagoura and Zakouas ont now integrated into the city.	-Extension areas -The area -The land use rate - The master plan -Number of districts -The infrastructure	- Field observation - Interviews with the Director MCLAU, the Mayor, the surveyor	Card Tables Graphics	**PART ONE:** THE DYNAMISM OF DALOA'S URBAN SPACE **Chapter 1:** The demographic dynamics of the city of Daloa **Chapter 2:** Spatial dynamics of the city of Daloa
The village communities and customary land rights holders of Sapia, Tagoura and Zakoua precede the subdivision of their plots of land to make them available to the city of Daloa.	How is land managed in the "integrated villages" of Sapia, Tagoura and Zakoua. What are "integrated villages"?	Studying land management in the "integrated villages" of Sapia, Tagoura and Zakoua.	Land production in the villages The integration of Sapia, Tagoura and Zakoua is achieved through internal organisation and subdivision operations.	Roledescheffe ries and Management committees - Subdivision procedure - Lot sharing - Lot size and price - Leprofilsocio-professi onal acquirers - The sale of land	-Interviews with traditional chieftaincies, -Interviews with the Director MCLAU, the Mayor, the surveyor, the Technical Director of AISCI -Survey by questionnaire	Tables Graphics	**PART TWO:** LAND MANAGEMENT IN INTEGRATED VILLAGES **Chapter 3:** Organisati on land in Sapia, Tagoura and Zakoua **Chapter 4:** Land production in the villages integrated
The subdivision of customary plots and the sale of land leads to land disputes that weaken the social cohesion. As a result of these disputes, the plots are insufficiently developed, thus making the urban landscape ugly (scrub, wasteland, sometimes used as a (e.g. intra-urban agriculture).	What is the impact socio-economic and spatial-environmental the urban dynamics ofDaloas on the land management in integrated villages Sapia, Tagourae t Zakoua?	Analysing the impact socio-economic andspati o-environmental dynamic s Daloa urban area onlagesti on land in integrated villages	Land speculation is at the root of land disputes and the rise of undeveloped plots that are marring the urban landscape of Daloa.	- Types of land disputes - Frequency of disputes - Actors in conflict - Modes of conflict resolution -The sale of land -The living environment -The typology of the habitat -The roads and utilities - property taxes and duties - equipme nt in infrastructure	-Field observation -Interviews with the Director MCLAU, the Mayor , the traditional chieftaincies, the expertgeomet er , theDire ctor AISCI technique. -Survey by questionnaire	Tables Graphics Photos	**PART THREE:** THE IMPACT OF URBAN DYNAMICS ON MANAGEM ENT LAND IN INTEGRATED VILLAGES **Chapter 5** The socio-economic impact of urban dynamics on land management in integrated villages **Chapter 6:** The spatial and environmental impact of the urban dynamics at financial managem ent in the integrated villages

THE DYNAMISM OF DALOA'S URBAN SPACE

24

Urbanisation in Cote d'Ivoire is characterised by the settlement and l^talement of cities in дёпёral, and particularly of regional cities. Daloa, capital of the Haut- Sassandra region is not on the fringe of this national trend. The demographic dynamics that the city of Daloa has experienced since the installation of the colonial military post in 1905, has an impact on the urban space that is undergoing changes.

In this first part of our study, the aim is to show how the dëmographic and spatial dynamics of the city of Daloa operate.

CHAPTER 1: THE DEMOGRAPHIC DYNAMICS OF THE CITY OF DALOA

Introduction

Demographic dynamics are a major component of the Involution of Territories. The interactions between man and his space contribute to its shaping and constitute an essential aspect of the geographical object. The interest of this chapter is to analyse the process of population settlement in Daloa.

1.1. Exponential population growth from 1921 to 2014

Due to its location at the crossroads of international roads, the city of Daloa has attracted large populations ëtrangëres to the region. Although declining, the population growth rate of the city of Daloa remains ëкуë: from 100.4% over the përiod 1975-1988, the growth rate increased to 42.1% between 1988 and 1998, then to 53.7% between 1998 and 2014 with respectively 60,800 inhabitants in 1975; 121,842 inhabitants in 1988, 173,107 inhabitants in 1998 and 266,000 inhabitants in 2014 (RGP 1975; RGPH 1988, 1998 and 2014). Figure 3 presents a revolution of the population of Daloa from 1921 to 2014.

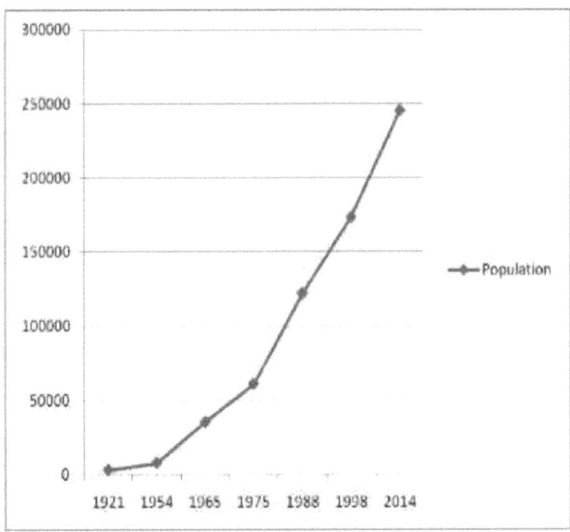

Source: Ecoloc 2002, RGP 1975, RGPH 1988, 1998 and 2014
Figure 3: Evolution of the population of Daloa from 1921 to 2014

The analysis of Figure 3 rëyëк that the population of Daloa grows exponentially over the entire përiod from 1921 to 2014. This population increase follows three trends: a relatively low growth from 1921 to 1975, a strong growth from 1975 to 1998, and a strong and abrupt population growth from 1998 onwards.

1.2. The stages of the demographic dynamics of Daloa

1.2.1. Remarkable population growth between 1921 and 1975

The population of Daloa grew at a low rate between 1921 and 1975. Table 7 presents the population trends of Daloa from 1921 to 1975.

Table 7: Evolution of the population of Daloa from 1921 to 1975

Years	Urban population	Rate of growth
1921	2811	3 %
1954	7487	
1965	35 000	5,7 %
1975	60 800	

Source: Ecoloc 2002, RGP 1975

The population of Daloa rose from 2800 in 1921 to 7487 in 1954, an average annual growth rate of 3%. This growth is low compared to the post-colonial period when the population was estimated at 35,000 in 1965, rising to 60,800 in 1975, an increase of 5.7%. In ten years, the demographic growth of Daloa has been sustained with an additional population of 25,800 inhabitants.

The settlement dynamics of the present site of the city of Daloa give rise to various theories. The indigenous people are thought to be the gban (gagou), bete and gouro peoples who inhabited the area before the plantation economy began in 1950. The bëtë are the majority group representing more than half of the indigenous population of the Daloa department (Ecoloc, 2002). The origin of the bëtë gives rise to very diverse hypotheses. Some believe that they come from Liberia (Delafosse, quoted by Diabate, 1987), while others find an endogenous origin (Louhoy, 1969). The bëtë appear to have been fixed since the end of the Neolithic between Bandama and Sassandra (Loucou, 1984). The Daloa beasts have Dri Kpekpa Dalo as their presumed ancestor (Zunon, 1980 cited by Alla D, 1991). He came from the Yukolu region in the south to hunt in the present Daloa region.

The demographic growth of the city of Daloa began in 1921. In 1921, the population was estimated at 2800 inhabitants. In 1954, 33 years later, Daloa had a population of 7500 inhabitants. The average annual growth rate over the period 1921-1954 is relatively low, around 3%.

Between 1954 and 1965, the settlement of the city accelerated with a growth rate of 15%. This demographic dynamic is influenced by the administrative function of the city. Indeed, from the strategic and military function conferred by the installation of the colonial military post in 1905, Daloa became the chief town of the administrative subdivision of the cercle. The improvement works of the city undertaken by Governor Peraldi in 1940 were an attraction for many people. Governor Peraldi drew up a plan for the development of the town and urbanisation became more and more marked. Daloa's position as a pole of local development is also affirmed by its status. Indeed, Daloa was established as a medium-sized commune with a municipal council elected on 18 September 1953. Then, Daloa became the capital of the Western Department following the decree creating the first four (04) departments of the country on 28 March 1959. This new administrative function of the city favoured the installation of an important population. In 1965, the 35,000 inhabitants, gave Daloa the third place in the urban hierarchy in Cote d'Ivoire, after Abidjan and Bouakë.

1.2.2. Strong demographic growth between 1975 and 1998

The population of Daloa grew at a high rate from 1975 to 1998. Table 8 shows the revolution of the population of Daloa from 1975 to 1998.

Table 8: Evolution of the population of Daloa from 1975 to 1998

Years	Urban population	Rate of growth
1975	60 800	4,6 %
1998	173 107	

Source: RGP 1975; RGPH 1998

Despite the decline in the growth rate to 4.   6%, the growth
The demographic growth of Daloa remains accentuated over the period 1975-1998. From 60,800 inhabitants in 1975, we have 173,107 inhabitants in 1998, which means an increase of 112,307 inhabitants.

This exponential demographic growth is the result of natural increase and strong migration towards the city. This situation attests to a rapid and marked growth, of the order of 10% per year (of which 8% is of migratory origin), which began at the beginning of the 1960s when Daloa was only an agglomeration of 18,000 inhabitants (YAPI DIAHOU, 1991, p. 1). The natural increase is due to a high fertility rate and a sharp drop in the mortality rate.

The demographic growth is favoured by the strengthening of the administrative function of the city. Indeed, Daloa was established as a full-fledged commune by law n°80-1180 relating to municipal organisation on 17 October 1980. Subsequently, Daloa became the capital of the Haut-Sassandra region by the dëcret n°96-567 of 28 August 1996, on the organisation of the national territory. Daloa has relatively better public services than the other departments of the country and concentrates many infrastructures. The populations of the region in particular, and that of the west of the country in дёпёral, prefer to settle in Daloa to bënëficier its collective services.

In addition, the town is experiencing an increasing number of migrants due to the development of the cocoa plantation economy in the region. Indeed, migrants who come in search of arable land for the creation of cocoa or coffee fields first settle in Daloa before infiltrating the region. The agro- ëcological diversity, the good availability of rain and surface water, and the great variety of arable land are the physical factors favourable to agricultural development in the region. At that time, the region ranked second in coffee and cocoa production with more than 20% of the national production. The strong regional attractiveness favoured by the development of cash crops resulted in waves of migration that continued to fuel the demographic growth of Daloa (Ecoloc, 2002).

In addition, the agglomeration of Daloa has an important influence on the surrounding rural environment and beyond, on a large part of the western half of the country. At the departmental level, there is a radiance of tracks converging towards the consumption centre of Daloa, which becomes a market where the surplus food production is devoured. On a national scale, Daloa is located in the most productive agricultural area of the country and benefits from road infrastructures that allow regular traffic of food products. Thus, located in the centre of an area of high agricultural production, Daloa appears as a "city in the countryside", constituting an immediate outlet for the region's food products and a source of supply for the entire Ivorian territory (Ecoloc, 2002).

1.2.3. Relatively low population growth between 1998 and 2014

The population of Daloa is growing at a relatively low rate between 1998 and 2014. The population growth rate, which was 4.6% between 1975 and 1998, fell to 2.7% between 1998 and 2014. However, there has been a considerable increase in the population. Indeed, over the period 1998-2014, Daloa recorded 92 893 additional inhabitants. Table 9 shows the evolution of the population of Daloa from 1998 to 2014

Table 9: Evolution of the population of Daloa from 1998 to 2014.

Years	Urban population	Rate of growth
1998	173 107	2,7 %
2014	245 360	

Source: RGPH 1998 and 2014

Between 1998 and 2014, the demographic growth of Daloa is favoured by the university function now acquired by the city. Indeed, the Unite Rëgionale d'Enseignement Supërieur (URES) of Daloa was created by dëcret n°96-613 of 09 August 1996. The URES of Daloa was created to concentrate the University of Abobo-Adjame and to reduce the important number of undergraduate students in Abidjan. The URES of Daloa has a training and research unit in Natural Sciences (UFR-SN) attached to the University of Abobo-Adjame. The mass of first year students in natural science of the University of Abobo-Adjame are now oriented to Daloa. The transformation of the URES of Daloa into the University Jean Lorougnon Guede by decree n°2012-986 of 10 October 2012, contributes enormously to the demographic growth of the city. In 2014, the University has more than 8000 students, 400 teachers-researchers and nearly 500 administrative and technical staff. Technical and vocational education is represented through the Centre de Formation Pedagogique (CAFOP), the Centre de Formation Professionnelle (CFP), the Centre Technique d'Apprentissage (CTA) and the Centre de Mecanisation Agricole (CMA).

In 1998, the urbanisation rate of the department is estimated at 35% and the city of Daloa hosts 33% of the departmental population (Ecoloc, 2002). In addition, within the framework of deconcentration, the city of Daloa hosts several regional directorates and branches representing almost all the administrative institutions in Côte d'Ivoire. Access to public services favours the settlement of a large population in the city.

From the 2000s onwards, a number of industries have been established in the city, contributing to its demographic growth. These industries are the main providers of employment and an essential component of the urban economy. These industries, which are essentially sawmills, are at the origin of the creation of an industrial zone and favour the arrival in the city of a massive population in search of employment as workers. In addition to this, there is the arrival of people who were displaced from their localities because of the military-political crisis in Côte d'Ivoire in 2002 and who settled there.

In sum, migratory movements have contributed to the demographic growth of Daloa. These populations were attracted by the functions of the city and by the development of trade due to the geographical position of the crossroads of the Centre-West region. However, the demographic growth of Daloa is not solely due to migratory movements, as natural increase contributes to it.

1.3. Densification of the urban space of Daloa
"The gëographer in the context of a population ëtude is more interested in the density of populations and the underlying explanatory factors. In the analysis of inhabited space, the geographer places in the foreground the phenomënes of population locations and distributions. The latter is concerned with the content and significance of settlement patterns. Similar crude densities have a different complementary geographical content" (DOLLFUS, 1973 quoted by KOUASSI, 2012).

1.3.3. Irregular evolution of general densities
"The general population densities in Daloa have evolved in an irregular manner, alternating between increases and decreases. The growth of densities induces a combined progression of densification and extension processes. The levels of density reached reflect the strong demand

of the populations in need of living spaces" (KOUKOUGNON, 2012).

Table 10 shows the evolution of population densities in Daloa from 1921 to 2014.

Table 10: Generative evolution of population density in Daloa from 1921 to 2014

Years	Population	Urbanised area (ha)	Density
1921	2811	152	18
1954	7487	240	31
1965	35 000	412	85
1975	60 800	838	72
1988	121 842	1964	62
1998	173 107	2592	67
2014	245 360	3191	77

Source: REPCI 1998, RGP 1975; RGPH 1988, 1998 and 2014

The analysis of the table shows that in general the population density in Daloa increased from 1921 to 2014. The evolution of densities has gone through three phases. From 1921 to 1965, there was a strong increase in density from 18 inhabitants/ha to 85 inhabitants/ha. Between 1965 and 1988, densities fell to 62 inhabitants/ha in 1988. From 1988 onwards, population densities in Daloa have been growing slowly.

From 1921 to 1965, the urbanisation of Daloa was achieved through settlement rather than through the expansion of urban space. Indeed, during the development of the plantation economy in 1950 and after independence in 1960, the privileged position of Daloa in the centre of an agricultural region attracted a large population. Major development works began and the scarcity of land favoured the densification of the first districts created.

From 1965 to 1988, the urbanisation of Daloa was carried out by extending the urban space. Numerous development works were undertaken because of the status of capital of the Centre-West. The emigration of people to Daloa led the authorities to open up numerous housing estates. From 1988 onwards, the growth in density corresponds to a simultaneous increase in population and urban space.

1.3.4 Unequal distribution of the urban population

The population is unevenly distributed in Daloa. Figure 4 shows the spatial distribution of the population density in Daloa.

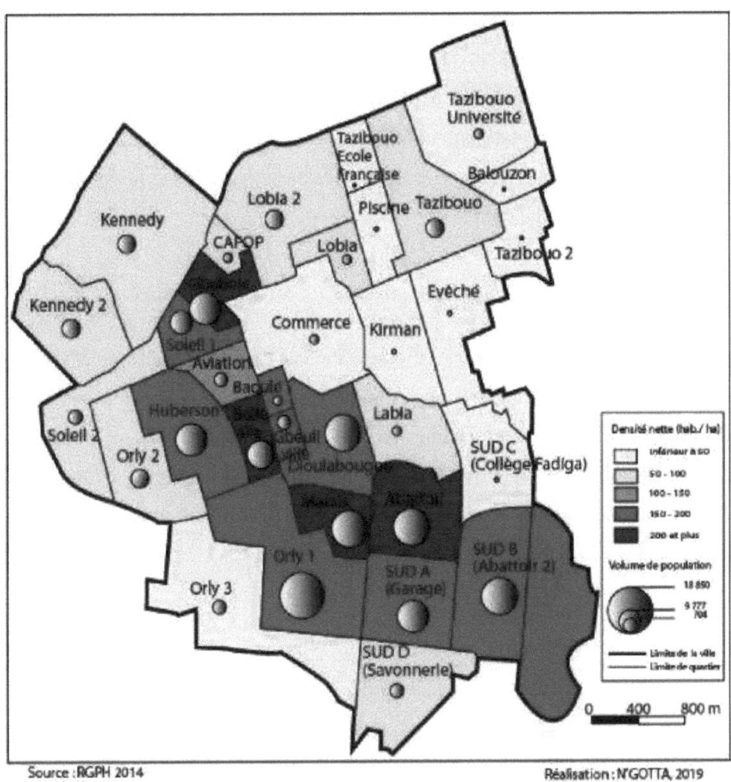

Figure 4: Population size and density of Daloa in 2014

The analysis of figure 4 reveals an uneven distribution of the population in Daloa. Four central districts, including Marais, Gbobek, Abattoir 1 and Belle-ville, concentrate 22% of the population in 12% of the territory. Population densities in Daloa increase as one moves away from the central districts (Marais, Gbobek, Abattoir 1 and Belle-ville). 24% of the population is concentrated on large areas (41% of the territory) in the central neighbourhoods Soleil 1, Huberson, Orly 1, Garage (South A) and Abattoir 2 (South B), where average population densities vary from 150 to 200 inhabitants per hectare. The lowest densities are recorded in the peripheral districts, especially in the eastern districts of Kirman, Eveche, Tazibouo 2, Tazibouo Ecole Frangaise, Piscine, Tazibouo Universe and South C.

The unequal distribution of population densities is Hëc' to the process of occupying urban space. The high population density observed in the central neighbourhoods Marais, Gbobek, Abattoir 1 and Belle-ville, which were the first neighbourhoods to be created, is explained by the scarcity of viable land. Subsequently, subdivision fronts were opened and made many lots available as the city developed. The lowest population densities are observed in the peripheral

31

neighbourhoods, which are residential, hence the low population numbers over a large area. The Commerce district, located in the centre of the city, has a low density because the population there is not residential. The population comes from all sides of the city to carry out commercial activities or to satisfy administrative, banking, pharmaceutical needs, etc. every morning and returns in the evening. Generally speaking, the population densities in Daloa decrease from the centre to the periphery of the city.

Conclusion

The population of Daloa has grown rapidly since the installation of the colonial military post in 1905. Population growth, which began weakly during the colonial period, has been sustained since the 1990s. The population growth is mainly due to the important migratory flows favoured by the administrative, military, university and commercial functions of the city. This population, which is also distributed over the urban area, contributes to its extension.

CHAPTER 2: THE SPATIAL DYNAMICS OF THE CITY OF DALOA

Introduction

The evolution of urban activities inevitably leads to a ëvolution of space consumption (FOFANA B., 2015, p. 71). The urban space of Daloa has grown since the installation of the colonial military post in 1905. The purpose of this chapter is to describe the different ëtages of the spatial dynamics of Daloa.

2.1. The stages of spatial growth of WAD

The spatial expansion of Daloa is marked by three main stages: the period from 1905 to 1980, the period from 1980 to 2000 and the period from 2000 to 2016.

2.1.1. Planned spatial growth before 1980

Prior to 1980, the spatial growth of Daloa followed master plans. From 1905 to 1980, Daloa had a planned spatial expansion: the land use followed the first subdivision plan and the first master plan of the city, respectively approved in 1929 and 1962. The town of Daloa was created with the installation of the colonial military post in 1905, in the centre of four core villages: Lobia, Labia, Tazibouo and Gbeuliville. These core villages are the original districts from which the town of Daloa was formed. Figure 5 shows the city of Daloa in 1940.

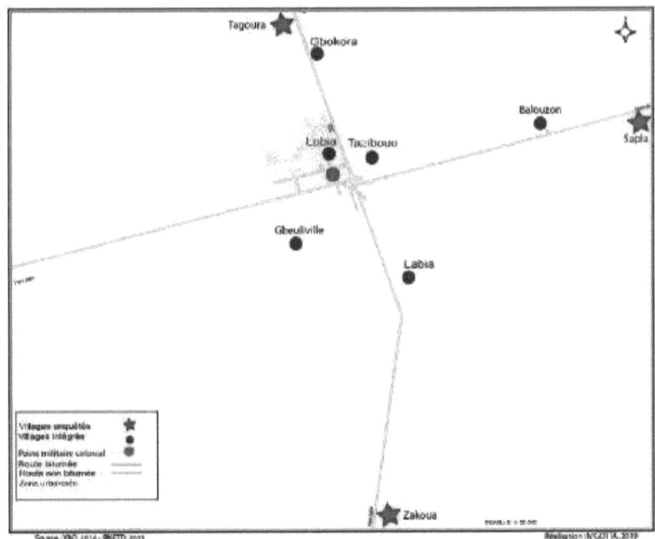

Figure 5: The city of Daloa in 1940

In 1940, the city was a patch around the military post in the Lobia 1 district. From 1940 to 1960, spatial growth moved southwards with the creation of the colonial district called

Commerce, which became the administrative and commercial district of the city. Dioulabougou, the so-called "African" quarter, was then created by Malinkës traders who settled around the colonial post as soon as it was created. The district attracted other communities from the sub-region who grouped together according to their origins (Malians, Guineans, Senegalese, Burkinabe), thus creating the Wolof, Segou, Cissoko, Mossibougou sub-districts. The saturation of the "African" district led to the creation of the Baoule and Gbeuliville popular districts. In 1959, the popular Marais district was created in the low-lying area, southwest of the city. The city expanded eastwards with the installation of the Centre Hospitalier Regional (CHR) and the Kirman Catholic College (see figure 6).

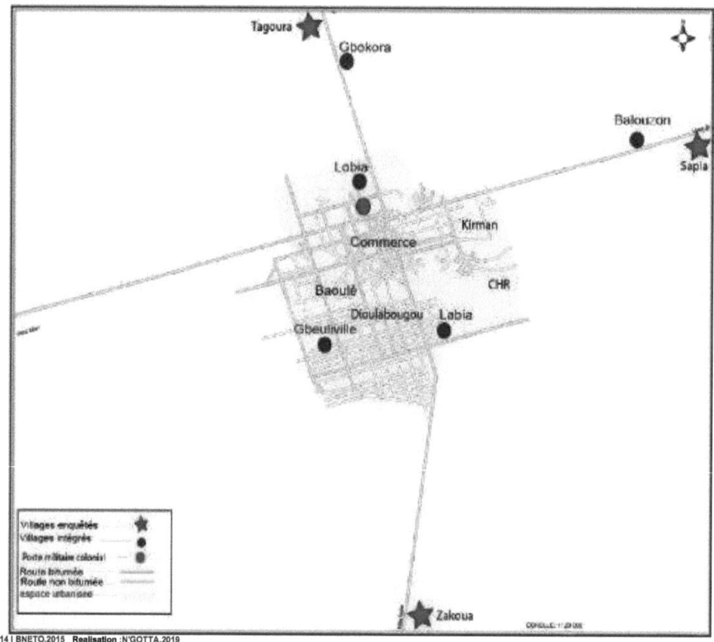

Figure 6: The city of Daloa in 1960

During the 1960s, the popular neighbourhoods Soleil 1, Aviation, Belleville, Gbobele and Kennedy 1 were created in the west of the city; in the north-east, the residential neighbourhoods Tazibouo 1 and Tazibouo Piscine were inaugurated in 1967.

During the period 1970-1975, the residential district Orly 1 was created in the southwest; the popular districts Abattoir 1 and Sud A in the southeast. In 1974, the spontaneous district Huberson, named after the prefect of the time, was created on the hillside to the west (see figure 7).

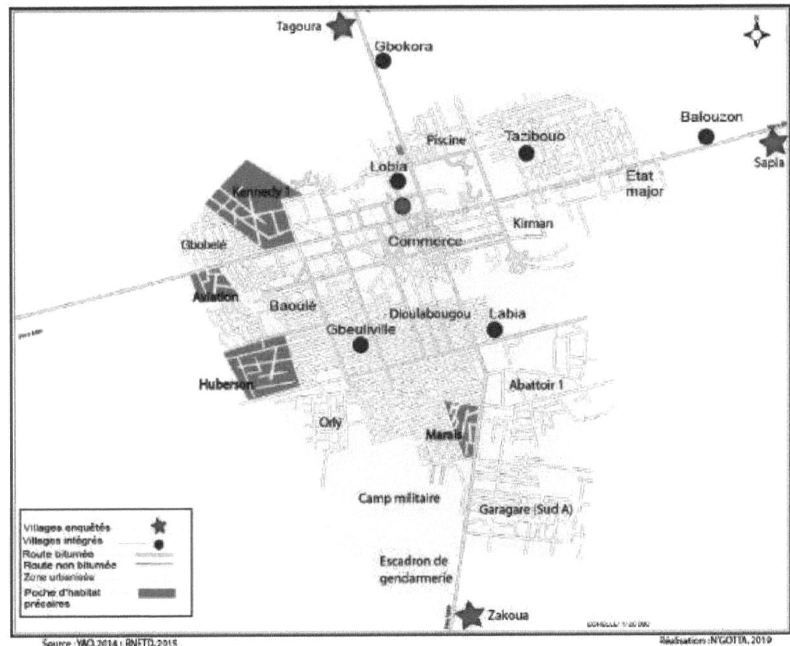

Figure 7: The city of Daloa in 1975

Between 1975 and 1980, the national gendarmerie was established, creating the headquarters, which occupies 45 hectares, and the squadron, which occupies 15 hectares, respectively at the entrance to the town on the road to Abidjan and at the exit on the road to Issia. The camp of the second military infantry battalion will also be created, juxtaposed to the gendarmerie squadron, to the south of the town. The village districts of Tazibouo, Labia and Lobia were restructured in 1980.

2.1.2. Proliferation of informal settlements between 1980 and 2000

During the period 1980 to 2000, we witnessed the proliferation of spontaneously created neighbourhoods on the margins of the city: this was the phase of the anarchic sprawl of Daloa.

The creation of spontaneous neighbourhoods in Daloa began with the country's independence and became more pronounced from 1980 onwards, the year in which the first municipal teams of the full-fledged communes were installed. Administrative slowness and the eagerness of candidates to build a home led to the spontaneous occupation of Daloa's urban land, which lasted over time (YAO, 2014). The administrative slowness has resulted in a shortage of building land and housing. This double deficit is one of the reasons for the development of unplanned housing (YAPI DIAHOU A., 1991, p113). In fact, from 1980 onwards, Daloa experienced a massive influx of people, which put pressure on the buildable space. The demand for land was then greater than the supply from the municipal services. However, the Mairie is confronted with the reluctance of customary owners who are rarely compensated after the subdivision of their plots. In the rush, people settle spontaneously on the margins of the city and other sites unsuitable for habitation (see Figure 8).

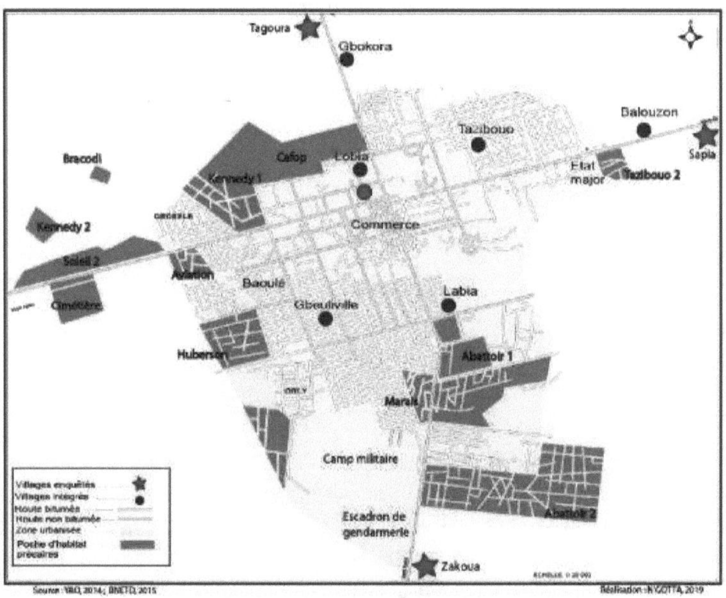

Figure 8: The city of Daloa in 1980

YAO (2014) explains that the spontaneous settlement of populations on the urban soil has led to the creation of precarious neighbourhoods on the outskirts of the city. Thus, the spontaneous neighbourhoods Abattoir 2 and Tazibouo 2 were created respectively on the South-East and East peripheries of Daloa; Orly, Huberson, Aviation, Soleil 2, Cimetiere on the South-West and West peripheries of the city, and the neighbourhoods Cafop, Kennedy 1 and 2, Soleil 2 and Bracodi in the North-West of Daloa. These spontaneous neighbourhoods cover large areas and are in дёпёral juxtaposition to the allocated neighbourhoods. Originally, these spontaneous neighbourhoods ёtaient plantations de cafё et de cacao ou des zones non aedificandi telles que les bas-fonds et les servitudes occupes iliegalement. People settle spontaneously because they lack the financial means to buy a plot or in the hope that an eventual subdivision of the plot they illegally occupy would allow them to obtain permanent plots. The occupants mainly use wood for the construction of the houses and recycled materials such as used or plastic sheets for the roofs. The absence of real estate or land promotion operations by organisations specialising in these fields places these 'owners' in a strategic position, to the point of becoming agents with a formidable hold on the development of the city: the layout of extension zones, housing typology, structuring of neighbourhoods, construction or not of facilities, etc. (YAPI DIAHOU A., 1991, p114).

The unhealthy living environment, the lack of drainage systems, frequent flooding and the high prevalence of certain diseases such as malaria, diarrhoea, cholera, etc., have prompted the municipal authorities to open housing estates. Between 1980 and 2000, the subdivision operations represented 773.6 hectares, or 9834 lots with an average of 800 m2 per lot. The subdivision fronts are open in the south of the city, particularly in the districts of South A, South B, South D and South C. These districts accounted for 38.6% of the subdivisions, or 185 hectares. Subdivisions are continuing in the Orly extension I, II, III and IV neighbourhoods, which represent 25% of the subdivision operations carried out (Daloa Town Hall).

2.1.3. Restructuring operations and spatial extension of Daloa from 2000 to 2016

Daloa has expanded rapidly as a result of restructuring operations in spontaneously created neighbourhoods.

From 2000 to 2010, the City Council will carry out restructuring operations to give legal existence to the spontaneous districts of the city created between 1980 and 2000. The restructuring consisted of the subdivision of pockets of precarious housing. With the opening of roads and the respect of the dimensions of the lots, the occupants who had not obtained a lot on the initial sites were resettled on the allotments called "extension". These "extension" districts will encourage the rapid spread of Daloa. Indeed, the restructuring operations that have taken place have led the Town Hall to open up new housing estates to accommodate the evicted populations, thus creating new neighbourhoods.

Lobia 2 extension in the North-West, Abattoir 2 extension in the South-East and Gbokora extension will appear on the city map. In the same logic, it is the creation of the Evëchë extension, Tazibouo 2 extension in the eastern part of Daloa; Sud D extension in the south and the Suisse district in the western part of Daloa. Orly 4, Orly 3, Orly Plateau will also be created in the South West.

Restructuring operations and the opening of new subdivision fronts to meet the demand for land have promoted the rapid expansion of the city (YAO, 2014). Table 11 presents the subdivisions built by the City Council from 2004 to 2007.

Table 11: Housing estates built by the Daloa City Council from 2004 to 2007

Neighbourhoods	Annedu	Number of	Lot size (m2)	Average area	Total area (m2)

	housing estate	lots		(m2)	
South/B Extension	2004	5185	300 a 600	450	2333250
Kennedy II Extension	2004	1900	300 a 600	450	855000
Tazibouo University	2004	1330	400 a 800	500	665000
Soleil II extension	2004	1037	400 a 600	550	570350
Tazibouo II North East		103	400 a 600	500	51500
Zakoua II	2006	377	400 a 600	500	188500
Eastern Corridor		277	400 a 600	500	138500
Green City		389	400 a 600	500	194500
Ex Aviation	2005	239	300 a 600	740	176860
Orly plateau	2007	921	400 a 600	500	460500
Olive trees		882	300 a 600	650	573300
Total		**12640**			**6207260**

Source: Daloa City Council, 2004-2007

From 2004 to 2007, the subdivisions developed have favoured the spatial expansion of Daloa. Over three (03) years, a total of 12,640 lots corresponding to a cumulative area of 6,207,260 m2, or 620.7 hectares, were made available. An average of 57 hectares are aménagës each year over this period. The lots are generally between 300 m2 and 800 m2 in size. The fronts of the housing estates open up in particular in three areas of the city. In the south of the city, the South B extension is the first sector concerned by the allotments with 5185 lots produced, equivalent to 233,250 m2 of urbanised area, i.e. 23 hectares (37% of the total allotments built). The development fronts then open up to the west of the city in the Kennedy II extension district. In this sector of the city, 855,000 m2 were developed, i.e. 85.5 hectares, providing 1,900 plots of between 300 and 600 m2 in size. The process of extension of the city is oriented to the north in the Tazibouo and Tazibouo Universite districts, where 1330 lots have been developed, corresponding to 665,000 m2 or 66.5 hectares; the size of the lots varies between 400 and 600 m2. The extension of the city continues to the west, in the Soleil 2 extension district. More than 1,037 lots with a total surface area of 570,350 m2 or 57 hectares are being developed. The size of the lots varies between 300 and 800 m2. The city has experienced throughout the period from 2004 to 2007 an urban development and an extension on its peripheries. In its extension, the city of Daloa consumes the space of the villages Balouzon, Bribouo, Gbokora, Sapia, Zakoua etc. which are today integrated into the city.

The city of Daloa now has 45 districts, and its expansion is ташёre continuous on its margins.

Figure 9 shows the stages of spatial growth in Daloa from 1905 to 2016.

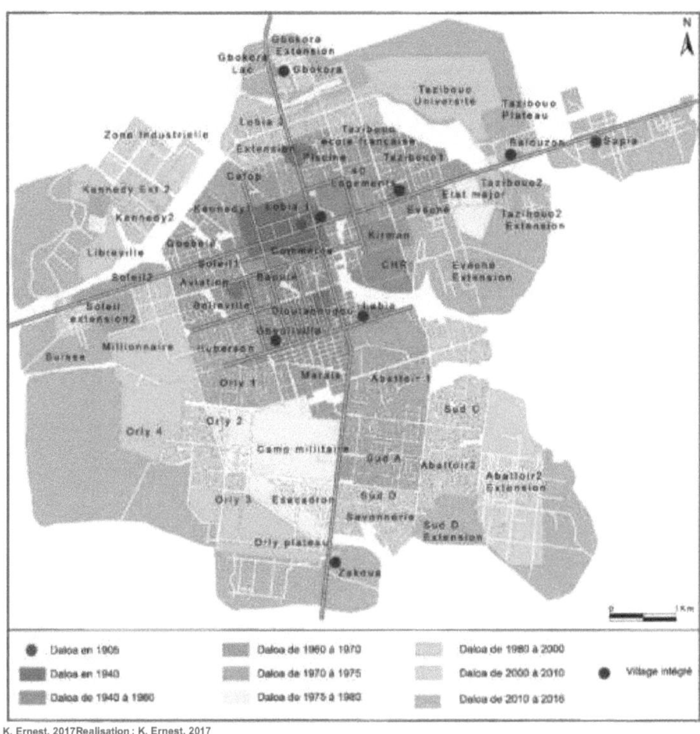

Investigations K. Ernest. 2017Realisation : K. Ernest. 2017

Figure 9: Stages of spatial growth in Daloa from 1905 to 2016

2.2. Evolution of the urbanised area of DALOA

The population growth of the city of Daloa has accompanied of pressure on urban and pēn-urbam spaces. From 242 hectares in 1958, then 393 ha in 1962, the urbanisë space successively passedë to 645 ha in 1970, to 1340 ha in 1980 then to 2510 ha in 1995. In 2008, the urbanisë space had reached 2,866 ha against 5500 ha in 2016 according to the technical services of the City Council (YAO, 2014). "Due to the dëmographic dynamics and the strong demand for building land, non-constructible spaces are taken by storm: the easements left by the amënageur around the lowlands, during the subdivisions, are convoked by the ever-growing population and other land seekers to build houses, putting to shame the prohibition of building in these areas" (ADOU G. et al., 2017, p. 64). Table 12 shows the evolution of the urbanised area of Daloa from 1958 to 2016.

Table 12: Urbanised area of Daloa from 1958 to 2016

Year	Urbanised area (ha)	Growth rate (%)
1958	242	
1962	393	12,88
1970	645	6,38
1975	838	5,37

1980	1340	9,84
1995	2510	13,37
2008	2866	1,02
2016	5500	2,01

Source: Ecoloc Daloa, 2002; Service technique de la Mairie, 2008, 2016

The analysis of the table rëyëк that from 1958 to 2016, the urbanisëe area of Daloa experienced a notable increase of 93% from 242 ha to 3300 ha. This growth is accentuated over the period 1970-1995 by 74% with 52% for the 1970s and 47% from 1980 to 1995. During the periods 1958 to 1970 and 1995 to 2016, the expansion of Daloa is less marked with respectively 62% and 24% growth. This spatial growth is represented by the following figure 10.

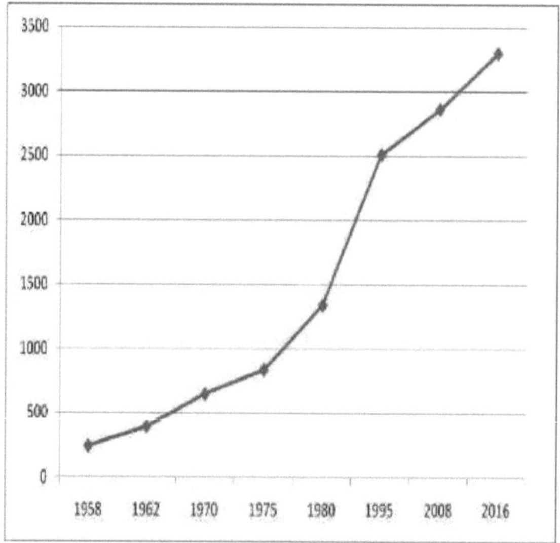

Source: Ecoloc Daloa, 2002; Service technique de la Mairie, 2008, 2016

Figure 10: Evolution of the urbanised area of Daloa from 1958 to 2016

The urbanised area of Daloa is growing exponentially and is marked by three main phases, namely the periods 1958-1975, 1975-1995, 1995-2016 where slow, abrupt and rapid, average growth is observed respectively.

From 1958 to 1975, the spatial growth of Daloa was slow, favoured by its administrative functions which attracted large populations. In fact, during this period, Daloa was successively a military post, the capital of an administrative subdivision of a circle and then of the Western Department, which included the towns of Man, Gagnoa, Bouafle, Issia and Vavoua. The city is equipped with administrative infrastructures requested by the populations of the region according to the strategy of development of the urban framework in Cote d'Ivoire. Between 1962 and 1975, out of 445 hectares of urbanisation, only 165 hectares, or 37%, had been foreseen by the plans; the other 380 hectares of urbanisation, or 63%, were the result of a spontaneous process (BCEOM, 1982).

During the period 1975-1995, the city began to grow rapidly, favoured by the acquisition of the function of military city and the strengthening of its administrative status. Indeed, Daloa is home to the second military infantry battalion and the headquarters of the national gendarmerie, whose military influence extends to the whole western half of the country. The military function confers a privileged position to the city of Daloa after Abidjan and Bouakd. According to the administrative division of 1990, Daloa became the capital of the Centre-West region, and by decree n°96-567 of 28 August 1996, on the organisation of the national territory, Daloa became the capital of the Haut-Sassandra region in 1996.

The spatial growth of Daloa is at an average rate from 1995 to 2016. This growth is impacted by the creation of the Regional Unit of Higher Education (URES) in 1996 by decree n°96-613 of 09 August 1996. The influx of students oriented to Daloa increases the demand for housing which favours the expansion of the city. Daloa became a university town in 2012 by the transformation of URES into Universitd Jean Lorougnon Guddd (UJLoG) through decree N°2012-986 of 10 October 2012. The institution promotes the arrival of more than 1000 students in the city each year. The demand for housing for students, teachers and administrative staff also contributes to the rapid spread of the city.

Furthermore, the installation of the UNOCI camp during the political and military crisis that the country has been experiencing since 2002 has contributed to the expansion of the urban area. The demand for housing by UN personnel has created a dynamic of house construction in order to rent them out. New buyers of urban land and owners of unfinished houses are busy finishing the construction of houses in order to put them on the "market". At the same time, the cost of renting houses for rent is increasing considerably throughout the city. The houses that have been built are mostly in the residential areas on the outskirts of the city.

The growth of Daloa is also influenced by the development of economic activities. Indeed, due to its position as a crossroads town, Daloa is a commercial pole where many products transit linking the North and South of the country; the countries of the hinterland Mali, Guinea, etc. to the port of San-Pedro; and the towns of the West of the country to those of the East. The city is also one of the main supply centres for food products in the whole of Côte d'Ivoire. Indeed, Daloa is the capital of a major food-producing agricultural region and constitutes a national market for the marketing of plantain, cassava, tarro, a ubergine etc.

Formed by four core villages, Lobia, Labia, Tazibouo and Gbeuliville, the town of Daloa has gradually spread out to its peripheries in the form of an aureole. The extension of the city has been achieved by incorporating the surrounding villages and consuming their land holdings. As a result of the sustained pace of spatial expansion, the villages of Balouzon, Bribouo, Gbokora, Sapia, Zakoua, Tagoura etc., which were once located at the periphery of the city, are now neighbourhood villages. Daloa reached Balouzon in 1970, Gbokora between 2000 and 2010, Sapia, Tagoura and Zakoua since 2016.

Daloa now has more than a hundred housing estates rëalisës on their land.

Conclusion

The spatial expansion of Daloa has been accentuated since the installation of the colonial military post in 1905. This expansion was done in three main stages. From 1905 to 1980, the city expanded according to master plans. From 1980 to 2000, the spatial growth of Daloa was not controlled and there was a proliferation of spontaneous (precarious) neighbourhoods within the city. The deficit in land production due to the slowness of the administration and the lack of financial means of the populations favoured the spontaneous occupation of the urban land of Daloa from 1980 onwards. From the 2000s, Daloa experienced rapid spatial growth through the restructuring of spontaneously created neighbourhoods. The restructuring operations will favour the creation of several new neighbourhoods called "extension" which accentuates the sprawl of Daloa. In its extension, the city of Daloa integrates the villages Sapia, Tagoura and Zakoua initially located in the periphery.

CONCLUSION OF PART ONE

The urban dynamics of Daloa, capital of the Haut Sassandra region, are remarkable. On the one hand, the settlement of the city, which started timidly during the colonial period, increased at a sustained rate from the 1980s. On the other hand, Daloa has experienced exponential spatial growth since the installation of the colonial military post in 1905, increasing from 242 ha in 1958 to 3,300 ha in 2016, i.e. an increase of 93% of its surface area. This spatial expansion has taken place on the land of the villages Balouzon, Bribouo, Gbokora, Sapia, Zakoua, Tagoura by integrating them. The dynamics of the city of Daloa is also due to its position as a crossroads city and its functions as an administrative, military, commercial and university city.

These data confirm our hypothesis 1 that

The remarkable spatial extension of Daloa, the peri-urban villages of Sapia, Tagoura and Zakoua are now integrated into the city.

PART TWO

LAND MANAGEMENT IN THE
"INTEGRATED VILLAGES" IN DALOA

Introduction

Land management is intrinsically Hëe to the development of territories. The access of the population to building land in the city implies the provision of rural land surrounding the cities. The process of transforming rural land used for agricultural purposes into lots (urban land) involves a land organization in the villages- intëgrës. In Daloa, the landowners of the integrated villages Sapia, Tagoura and Zakoua are initiating the subdivision of their plots. The question is: how is дëгë land in the "inlegjes villages" in Daloa?

In this part, Chapter 3 presents the land tenure organisation in the villages intëgrës by ëvoking the prëpondërant role of the chiefdoms and land management committees and the way land is shared.

Chapter 4 shows how subdivision operations in intëgrës villages are carried out and analyses subdivision production and lot development.

2.3. ITER 3: LAND TENURE IN SAPIA, TAGOURA AND ZAKOUA

Introduction

Having integrated their villages into the city, the customary landowners of Zakoua, Tagoura and Sapia are organising themselves to make their land available to the state. So how is the land organisation in these integrated villages?

This chapter is structured in two parts. In the first part, we will discuss the prëpondërant role of traditional chieftaincies and land management committees. The second part deals with land ownership in integrated villages through.

3.1. Land managers in integrated villages

The village chief and the notabilite, the land chief and the management committee are the land managers in the integrated villages SAPIA, TAGOURA and ZAKOUA.

3.1.1. The role of traditional chieftaincies

The traditional chieftaincies play a major role in land management in Sapia, Tagoura and Zakoua, particularly with regard to subdivision operations and the settlement of land disputes. They organise and supervise the subdivision operations in order to avoid anarchy in land management. The control is done through the establishment of a register in which all the subdivision operations of the village are recorded. This register is used to avoid multiple sales of land. In addition, the village chiefdoms have an arbitration function in land disputes. It is in their presence that disputes are settled, and their aim is to find agreements between the protagonists in order to promote peace in the village. The chief of Zakoua expresses this concern for cohesion in these words: *'The village chief is there to harmonise when there are disputes... and we always make sure that the villagers and the occupier are always in agreement. We have to avoid clashes.*

The chiefdom is also called upon by the administrative, police and judicial authorities to settle land disputes in their sectors. 70% of respondents said they had used the chiefdom to settle a land dispute. Figure 11 shows the distribution of recourse to the chiefdom in the event of a land dispute.

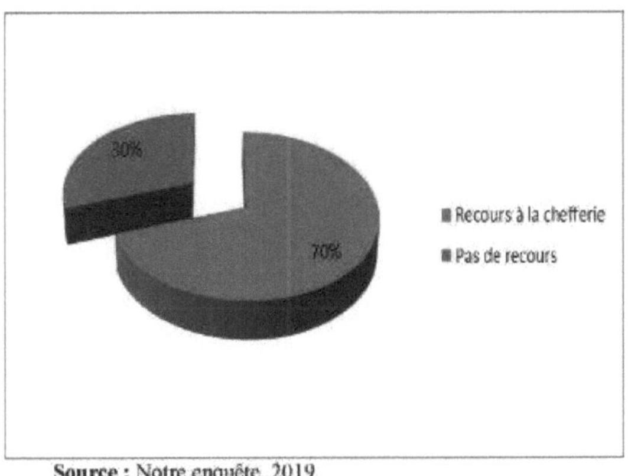

Source : Notre enquête, 2019

Figure 11: Distribution of recourse to the chiefdom in the event of a land dispute

Analysis of the graph reveals that almost three quarters (70%) of customary landowners have resorted to the chieftaincy for the settlement of land disputes. This demonstrates the important role of the chieftaincy, which is systematically called upon in the event of a dispute. In addition, the chiefdom is always notified before a village plot is subdivided.

3.1.2. The role of land management committees

In each of the three villages surveyed, there is a land management committee. This committee is composed of 'knowers' who know the boundaries of the village land and the boundaries of the village inhabitants' plots.

The intervention of the management committee takes place primarily at the level of subdivision operations for the delimitation of plots. The land management committee organises and supervises all subdivisions.

In addition, the land management committee is the first recourse in the event of a land dispute; it involves the chiefdom depending on the seriousness of the land dispute. The protagonists are obliged to respect the convocations of the president of the land management committee and its decisions are most often accepted because of the authority vested in it. The land management committee is constantly called upon by customary owners: 67% of the respondents said that they had used the committee during the parcelling out of land, and 33% in the event of a land dispute (Our survey, 2019). Figure 12 shows the distribution of reasons for using the land management committees.

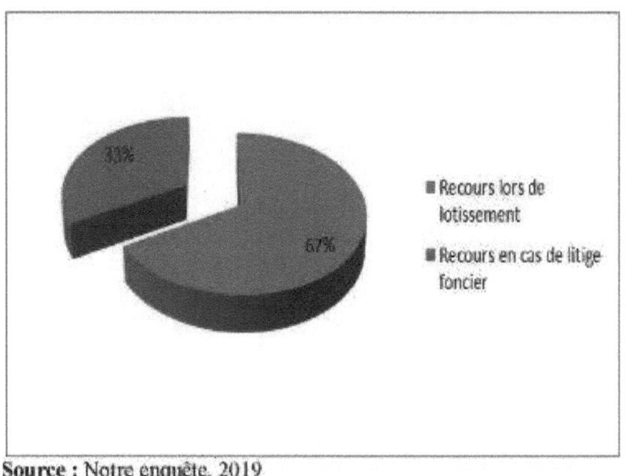

Source : Notre enquête, 2019

Figure 12: Distribution of reasons for using land management committees

The analysis of the graph shows that the соткё of land management intervenes primarily during the opёгайопз of subdivision (67%). They are solicitedёs to a lesser extent for the settlement of land disputes (33%). Indeed, in the exercise of their functions, the land management committees aim to prevent land disputes. This is why they are more often called upon to intervene upstream during the parcelling out of plots.

Speaking about the role of the traditional chieftaincy and the land management committee, the chief of Tagoura concludes his remarks by saying: *'the village chief is really at the heart of this management with his rural land committee'*. The land management committee plays a major role in the regulation of the land tenure system in Zakoua, Tagoura and Sapia.

3.1.3. The role of land managers

Land is a community asset in Sapia, Tagoura and Zakoua where management is entrusted to individuals who hold land powers.

In Sapia and Tagoura, land power is held by the chiefs of the lineages. The village is, in effect, an ent^ composed of several lineages. Each lineage has its land which is managed from father to son. The chief of Tagoura explains that: *'The chief of the lineage has all the land. He allocates it to the 'glegbons', i.e. the heads of the families who, in turn, allocate it to their children. So there is always someone to manage the customary land.*

Lineage chiefs are the exclusive holders of customary land and have the duty to distribute it within their families. The Secretary General of Sapia states that *'there is no central land chief in Sapia; each landowner is the head of his or her landholding. The head of the family is the head of the land"*. There are consequently as many landholders as there are heads of families within these villages. Each head of family has a piece of land that he manages on behalf of his family. In Tagoura, however, there is a land chief who oversees the management of the land.

In Zakoua, unlike the other two villages, land power is held by a single person called the land chief or "proprietaire terrien". This is due to the fact that the village is composed of a single indigenous Bëtë family. The land chief is ëШ by the heads of the family during a meeting to ensure the management of the family's land holdings.

3.2. Customary land ownership in integrated villages

Indigenous management of land differs from one locality to another; the relationship of populations to land being a cultural fact. Land is a heritage of the village community shared within families in a traditional way in which plots are well dëlimitëes.

3.2.1. Land: a family asset

The system of land sharing differs significantly from village to village.

In Tagoura, the head of the family distributes the land among all his male children. However, if there is a child who proves to be more valiant at work, he works on many more plots than his brothers; since the land belongs to the family, the bravest member is allowed to enlarge his field by granting him another portion.

In Sapia, the land is shared between the male children. The Secretary General clarifies the sharing method in these terms: "*We share the land between the male children. The girls have no right to the land because they will go and get married. Our parents had several wives, each one could have four, five, six wives, even more. The plot of land that each woman worked on, that's where her children are placed. Each child goes to his mother's side, under the control of the father.*" The beneficiaries of the land are distributed geographically over the family's land assets according to the position occupied by the mother; the area of each is determined by the number of children the mother has.

In Zakoua, sharing within the family is done by consensus in equal shares, from the largest to the smallest. However, Гзтё always has one more share. Today, plots of land are allocated to women who were previously excluded from sharing the land. In fact, according to custom, women are not taken into account in the sharing of land in the three villages. They are considered as inheritors of their husbands' wealth in the host family.

3.2.2. Recognition of boundaries between plots

The ëlëments for recognising the boundaries of the plots are identical in Sapia, Tagoura and Zakoua. They are easily identifiable natural features. The chief of Zakoua summarises these identifiers: '*They can be large trees located at the boundary of the plots. They serve future generations to recognise the boundaries of their parents' plots. To avoid conflicts, these trees are not cut down. Teak trees can even be planted at the boundary and everyone knows his or her plot. The boundaries of the plots are also symbolised by termite mounds or sometimes rivers.*

Furthermore, the recognition of boundaries can be done through the contribution of the 'knower' who can be the chief of the land or a family member. The chief of Tagoura says: *'The chief of the land can be a 'savoir'. But there are people who have been there since and who hardly move, because they are relays. When you have your father there and you are by his side, he has shown you the limits. But it's not in every family that everyone leaves to work elsewhere. There is always someone who looks after the clans' assets.* The family members who have chosen to stay in the village know the limits of the family land.

Conclusion

The traditional chieftaincies and the land management committees of the intëgrës villages organise and supervise the subdivision operations and participate in the settlement of land disputes. In these villages, the ownership and sharing of land is significantly different. The customary landholders of Sapia, Tagoura and Zakoua initiate private land allotments.

CHAPTER 4: LAND PRODUCTION IN 'INTEGRATED VILLAGES' IN DALOA

Introduction

Landowners are initiating private housing developments since the end of administrative housing developments. In Daloa, private subdivision operations are being carried out in the integrated villages of Sapia, Tagoura and Zakoua. This chapter analyses the production of subdivisions and the development of plots in the integrated villages.

4.1. The actors of land production in integrated villages

The actors in land production are divided into three categories: the public authority, the applicant and third parties (see Figure 13).

Source: Our survey, 2019
Figure 13: Actors in land production in integrated villages

The subdivision is an operation carried out by the public authority represented by the Ministry of Construction and Urban Planning, and the applicant, who may be the holder of customary land rights called the 'promoter', village communities and decentralised authorities. Local residents, companies and real estate agencies are the catalysts of these subdivision operations.

The Ministère de la Construction et de l'Urbanisme is in charge of the administrative management of subdivision files. It is the Ministry that receives the subdivision application at its one-stop land office and processes the subdivision file until its approval. The processing consists of the control of the conformity of the subdivision projects with the master plan of the city. In addition, the Ministry of Construction and Urban Planning issues laws on urban land and ensures that urban planning standards are met when houses are built.

Customary landowners, village communities and decentralised authorities (Mairie, Regional Council) are the applicants for subdivision. The customary landowner is the one who owns the plot of land to be subdivided. This property is acquired according to custom, in a traditional manner. In some cases, it is a person designated as a "representative" by the family. These "representatives" are usually the elders of the family, executives often called "intellectuals" or a member recognised as being of good character by the family. In addition, village communities can apply for housing estates for the extension of their villages. In addition, town halls initiate subdivision applications as project owners. According to YAO et al (2017, p. 512), there is an autochthonous-municipal complementarity in the production of urban space in Zoukougbeu (Cote d'Ivoire). Following the request of the village collectivities, the mayor's office undertakes to build housing estates for their benefit.

The demand for land (population, companies, real estate agencies, etc.) accelerates the pace of development: the high demand for urban land accelerates land production.

Figure 14 shows the stages of land production in the integrated villages.

Source: Our survey, 2019

Figure 14: Stages of land production in integrated villages

The analysis of Figure 14 shows that, in the integrated villages, land production is left to the initiative of the customary owner. The latter sends a request for the subdivision of his plot of land to the one-stop shop of the Regional Directorate of Construction and Urban Planning, which forwards it to the central directorate in Abidjan. In response, the central directorate requests a commodo et incommodo investigation in order to determine whether the plot to be subdivided is not subject to conflict. This investigation is carried out by the town hall if the plot to be developed is within the municipal boundaries, or by the sub-prefect if it is outside the municipal boundaries. If there is no opposition, the geomdtre proceeds to the polygonal which consists in delimiting the plot. The town planner then draws up the subdivision project by dividing the plot into lots according to the town's master plan. The subdivision project is sent to the central directorate of the Ministry of Construction and Urban Planning in Abidjan for the signature of the subdivision approval order by the Minister. The subdivision approval order marks the official existence of the lots on the plot. The subdivision file is raтeпë to l'urbaniste who entrusts the geometre with the demarcation.

4.2. The work agreement with the surveyor or the real estate agency

Our surveys reveal that 92% of landowners have their plots subdivided by surveyors and 8% by real estate agencies. This disproportion is due to the fact that surveyors visit the villages to motivate landowners to have their plots subdivided. They offer to carry out the subdivision operations in return for remuneration, usually in kind, as the customary owners are unable to pay them in cash. A work agreement is signed between the landowner and the surveyor or real estate agency in charge of the subdivision, which sets out the obligations of each party.

The obligations of the landowner are as follows:

- Inform the competent authorities in the town (the Prefect, the Sub-prefect, the Regional Director of Construction and Urban Planning, the Mayor and the village chief) of the subdivision work on his plot;

- To solve all the problems in case he has opposition from his peers, from an administrative or private structure;

- To compensate the project owner if any conflicts of ownership should disrupt or stop the work for any reason;

- Provide 50-60% of the lots needed to open the lanes.

The obligations of the surveyor or the real estate agency financing the development, referred to as the "project owner", are set out below:

- Boundary of the plot to be developed;

- Realise the housing estate project;

- Follow the subdivision project until it is approved;

- Realising the housing estate ;
- Establish and produce the final technical file;
- Provide 35% of the lots needed to open the lanes.

In addition to these provisions laying down the general framework of the work, 60% of the plots realised go back to the landowner for the purging of his customary rights. In the case of the realisation of a public сПпПИё project by the State or a private developer on the plot, five percent (5%) of the amounts to be paid are granted to each farmer for the purging of the cultivation rights. In the framework of the subdivision, ten (10%) to twenty (20%) percent of the sixty (60%) percent of the lots to be built constitute the purging of the landowner's soil rights and the remaining fifty (50%) to forty (40%) percent constitute the farmer's allowance for the purging of his cultivation rights. The maximum cost of purging for the loss of customary rights throughout the national territory varies between six hundred (600) CFA francs and two thousand (2000) CFA francs per square metre, according to Decree No. 2014-25 of 22 January 2014 amending Decree No. 2013-224 of 22 March 2013 regulating the purging of customary land rights for general intdict. The subdivision works shall be carried out in accordance with the rules and standards in force contained in the technical instructions of the Ministry of Construction and Urbanism.

4.3. The division of lots at the end of the subdivision

At the end of the allotment, the lots are divided between the landowner, the surveyor or the real estate agency and the administration. Table 13 shows the distribution of lots in a private housing estate in Sapia.

Table 13: Distribution of lots in a private housing estate in Sapia

Awardees	Number of lots awarded
Mr KANON Guero David	134
A.I.S.C.I	65
Administration	24
Opening of the tracks	20

Source: Our survey, 2019

The subdivision was built on a 16 hectare plot of land owned by Mr. KANON Guero David. Financed by the real estate company AISCI, the subdivision of KANON D.'s plot has made 243 lots available. Out of the 243 plots, 134 plots (55% of the total) are owned by the customary landowner, 26% of the plots are owned by the project owner (AISCI) and 10% by the administration (Town Hall, Regional Council, Ministry of Construction and Urbanism). The road system occupies only 8% of the allotted plot. In fact, lots are used for the opening of roads when the urban planning project is established by the urban planner. The division of the lots is done according to the distribution key agreed upon in the work agreement signed between the customary owner and the project owner. Each of the parties is free to use its lots. Moreover, the share of the lots allocated to Mr. KANON are shared by the family in case he is the representative of the family. The lots that belong to the administration are accumulated in the form of administrative reserves. After the division of the lots, a report is drawn up.

A Proces-Verbal de partage des lots is signed at the end of the realization of the approved allotments at the initiative of the village communities. The Proces-Verbal de partage des lots (PV de partage des lots) determines the list of bënëficiaries of the lots and constitutes a beginning of proof of right according to article 2 of the arrete. The lots are shared by consensus between the actors involved in the realization of the allotment, namely the landowner, the geomëtre having carried out the demarcation of the plot or the real estate agency having financed the allotment operation, the administration, and the village community. The share of the gëomëtre dëtre depends on the work agreement signed with the landowner. It is optional whether he is paid in kind or in cash. However, the surveyor is generally paid in kind for three (03) lots out of ten (10) carried out. According to the provisions of article 3 of the decree, the Proces-Verbal of the division of the lots is drawn up under the supervision of a sworn officer of the Ministry of Construction and Urbanism in order to draw up the Attestation Domaniale. The original copy of the Proces-Verbal is sent to the officials of the decentralized services of the Ministry of Construction and Urbanism as well as to the regional prefect. The division of the lots can be established by a bailiff.

4.4. The development of plots in the "integrated villages" in DALOA

4.4.1. The procedure for purchasing lots in villages integrated into DALOA

The purchase of periurban land in Zakoua, Tagoura and Sapia follows a procedure in which the chiefdom intervenes.

In Tagoura, a certificate of sale is drawn up and signed by the landowner, the purchaser, and their witnesses. The village chief signs it after a check in the register where all the subdivisions made are recorded to verify that the land has not already been sold. Then, the certificate of sale is sent to the office of the Association des Proprietaires Terriens de la Commune de Daloa (APTDC) where a village certificate of transfer is issued at 50,000 CFA francs. The certificate of village transfer is signed by the president of the association, the village chief and the purchaser, to certify the customary recognition of the property. With this attestation, the purchaser can finally have his Attestation de Concession Definitive (ACD) issued by the Ministry of Construction and Urbanism.

In Sapia and Zakoua, on the other hand, a village attestation is signed by the landowner, the landowner and their witnesses, and the village chief. Only the village chief issues the village certificate, which is paid for by the purchaser. The price of the village certificate varies from 50,000 to 70,000 CFA francs, of which 30,000 CFA francs is paid into the commune's landowners' association. Sometimes the amount of the village certificate is included in the cost of the lot. In this case, the purchaser receives the village certificate issued by the village chief from the landowner after paying the lot price. The village attestation is part of the documentation required to obtain the CDA. It is a prerequisite.

The purchase of land from landowners is secure and avoids cases of selling a plot to several people. The purchaser has the possibility to check whether the lot has not already been allocated to another person. This verification is possible at the village registers, at the office of the Association des Proprietaires Terriens de la Commune de Daloa (APTDC), at the regional directorate of the Ministry of Construction and Urbanism and at the cadastral department. In addition, the procedure for purchasing peri-urban land is fast and without any waiting period between the different stages. However, the lack of a formal framework for land publicity makes it difficult to purchase peri-urban land. Indeed, applicants for land purchase go to the villages to meet landowners willing to sell plots. Sometimes it is the brokers who act as intermediaries between the applicants and the landowners. In this case, the price of the land is usually high because the landowner pays the agent 10% of the cost of the lot. In addition, the location of the proposed plots of land is not always suitable for the land applicants.

The office of the Daloa-Commune Landowners' Association (photo 1) is the place where the village land transfer certificate is issued. It also serves as a place for settling land disputes and as a forum for discussion among landowners in the commune.

Photo 1: Office of the landowners' association of the commune of Daloa

4.4.2. Lot sizes and prices

The size and prices of the lots differ from one village to another. Table 14 shows the size and purchase price of the lots рёгшгҍат.

Table 14: Size and purchase price of peri-urban lots

Villages	Lot size	Purchase price of the lots (F
Sapia	600 m^2 and 500 m^2	from 300 000 to 2 000 000
Tagoura	400 m^2 and 600 m^2	250 000 a 4 000 000
Zakoua	600 m^2	300 000 a 1 000 000

Source: Our survey, 2019

In Tagoura, the lots are 600m^2 and 500m^2 and have prices that vary from 300,000 to 2,000,000 FCFA depending on whether the land is located near the primary asphalt road or not. The price of the lots is also ёпАиспсё by the proximik of the national gendarmerie training school located in Toroguhe; the acquirers feeling more sёcuritё.

In Sapia the size of the plots varies between 400 m^2 and 600m^2 . On the same plot, one can find lots of 400 m^2 , 600 m^2 and a little more. The price of the land depends on the distance to the village and ranges from 250,000 to 4,000,000 CFA francs.

In Zakoua, the plots are all rectangular, 30 m long and 20 m wide, i.e. 600 m^2 . The prices of the lots vary from 300,000 to 1,000,000 CFA francs.

Our survey reveals that landowners do not wait for their land to be serviced before selling it, which is the reason why prices vary from place to place. Indeed, some of the lots purchased correspond to land on which coffee and cocoa plantations still exist. The cost of these plots is less ёкуё because the buyer cannot proceed to the construction of houses in Птёе&a! These plots are in дёпёга! ёloignёs from the village. Moreover, the purchasers deprive the plots on flat and easily accessible surfaces at the expense of sloping or elevated plots.

The variation in land prices creates a dёsordre in the пки'ёЬё foncier pёriurbain which is not always to the advantage of landowners. Indeed, customary landowners very often sell their lots at derisory prices because they are subject to the law of supply and demand. For juxtaposed lots belonging to two landowners, one facing a dёpense, usually a case of family dёcёs, can sell his lot at 300 000 FCFA, while the neighbouring landowner offers his lot at 1 000 000 FCFA. In this situation, the second landowner ends up cёdering and also sells his lot at a price of around 300,000 CFA francs to the bёrайec of the buyer. In the opinion of the village chief of Tagoura, '*in principle, the prices of these lands should be discussed within the landowners' association and a uniform price should be set. Unfortunately, this is not the case and prices are made in different ways.* The Tagoura chief's comments thus highlight the action of the association of landowners in the commune, which could serve as a framework for discussion and standardisation of land prices, thereby avoiding the disorder observed in the land market in Daloa. In Daloa, the përiurban plots are in дёгага! large sizes (600 m2) and prices vary from 250,000 to 4,000,000 FCA.

The price of përiurbains is also influenced ë by the lack of a formal framework for public^. Each landowner has the latitude to set the price of his land. The landowner then offers his land at a price ёкyё so as to have a better profit after this price is settled. Figure 15 presents the pricing system for urban land.

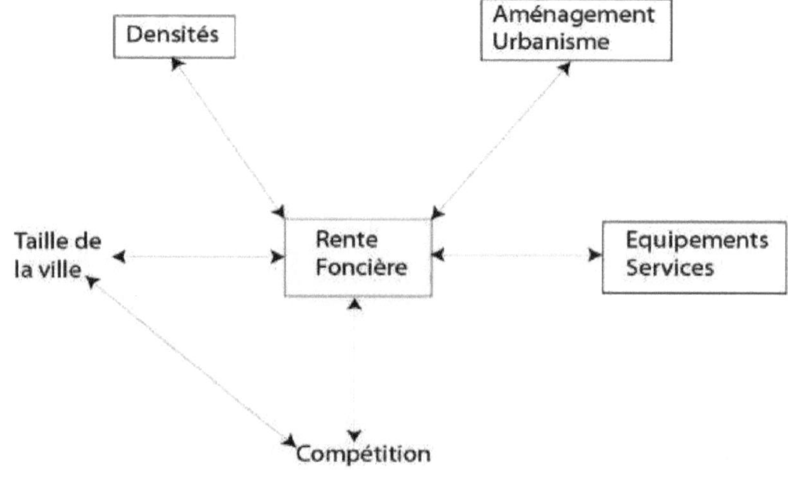

Source: Our survey, 2019
Figure 15: Urban land pricing system

At the heart of the system is land rent, a dependent variable influenced by various variables, including the layout of the urban space, the provision of infrastructure and the development of services, densities, the size of the city and the distribution of actors for land production. In this open system, a cycle is formed by the variables city size, the distribution of actors and land rent. Land rent comes under increasing pressure as the city develops. The price of land in the peri-urban areas is increasingly high with their integration into the city and fluctuates according to the type of neighbourhood (residential or popular) and the presence of the facilities located there.

Conclusion

Private subdivisions initiated by landowners are based on a work contract signed between the landowner and the surveyor or real estate agency. At the end of the subdivisions, the landowners proceed to the sale of their share of the lots with a village certificate issued by the village chief.

CONCLUSION OF PART TWO

In Daloa, village chiefdoms and land management committees organise and supervise the subdivision of land in integrated villages. For the subdivision of their portions of land, the landowners solicit a surveyor or a real estate agency who is paid in kind. At the end of the subdivision, the lots are divided up according to the previously signed work contract. Each party then proceeds to sell its share of lots.

Therefore, hypothesis 2 of the study, which states that 'land production in integrated villages is carried out through internal village organisation and subdivision operations', is confirmed.

What is the socio-economic and spatio-environmental impact of the urban dynamics of Daloa on land management in the integrated villages of Sapia, Tagoura and Zakoua?

PART THREE

THE SOCIO-ECONOMIC IMPACT AND THE SPATIAL-ENVIRONMENTAL IMPACT OF DALOA'S URBAN DYNAMICS ON LAND MANAGEMENT IN INTEGRATED VILLAGES

Introduction

Urban dynamics have an impact on land management in integrated villages. In Daloa, this impact is manifested at the social, economic, spatial and environmental levels. The purpose of this section is to analyse the socio-economic and spatial-environmental impact of Daloa's urban dynamics on land management in the integrated villages.

Chapter 5 shows the socio-economic impact of urban dynamics on land management in integrated villages by evoking the fragility of social cohesion, the non-respect of traditional chieftaincies, the multiplication of land disputes on the one hand; and the ëconomic impact through the analysis of spëculation Гопйёге, the payment of land taxes, economic reconversions in integrated villages, on the other hand.

Chapter 6 analyses the spatio-environmental impact of urban dynamics on land management in integrated villages, including the proliferation of undeveloped lots, the development of intra-urban agriculture, etc.

CHAPTER 5: THE SOCIO-ECONOMIC IMPACT OF DALOA'S URBAN DYNAMICS ON LAND MANAGEMENT IN INTEGRATED VILLAGES

Introduction

Controlling the rapid expansion of urban space is a ёёй that mayors and other managers of Ivorian cities are facing. In Cote d'Ivoire, urban land production is a market subject to strong spëculations since 2014. This chapter analyses the socio-economic impact of Daloa's urban dynamics on land management in the integrated villages Sapia, Tagoura and Zakoua.

5.1. The social impact of urban dynamics on land management in integrated villages

Land management in Ыёдхёз villages is accompanied by conflicts that weaken social cohesion and lead to the disrespect of traditional chieftaincies.

5.1.1. The fragility of social cohesion

In the integrated villages of Sapia, Tagoura and Zakoua, land has always been a community asset managed by the family. Nowadays, possession of a plot of land with well-defined boundaries is a stake that is not to be negotiated. This stake often creates claims, and encroachments can degenerate and be a source of division within families or even clans. The Secretary General of Sapia rightly says: "*When you have your property and someone wants to take it away from you, if you are convinced that it is your property, then you do not accept it. And usually people get into situations of no return, where they can go as far as locking the other person up, or they can go as far as fighting. And all this can create uncomfortable situations.*"

For the villagers, land is the most precious asset bequeathed by their ancestors, more important than gold in the imagination of the villagers, which must be preserved and safeguarded at the cost of one's life if necessary. Conflicts over land often create divisions within families and village communities. Figure 16 shows the extent of conflict within families during the process of land division.

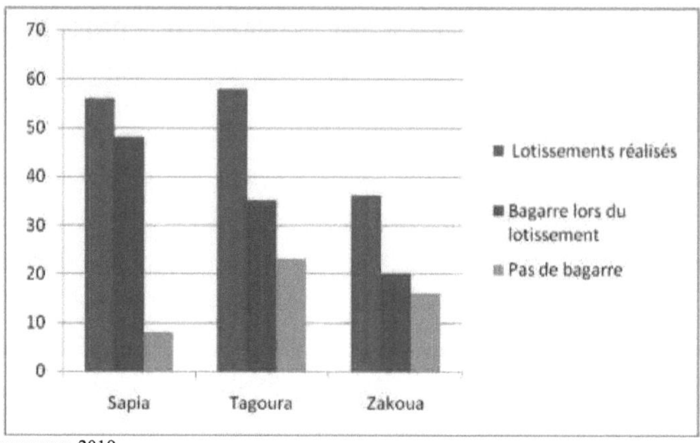

Source: Our survey, 2019

Figure 16: Conflict within families during housing developments

The analysis in Figure 16 shows the dominance of fighting at the time of subdivision (69%) in the integrated villages. These fights are important in Sapia (85% of the customary owners of the village are concerned), Tagoura (60%) and Zakoua (55%). The scale of the fights in Sapia, during the housing developments, can be explained by the strong pressure and speculation over land. Indeed, the privileged position of Sapia on the national road to Abidjan accentuates the pace of subdivisions. The land rent in Sapia is subject to increasing fluctuations because 'everyone wants to have their house on the road to Abidjan'. The growing speculation is the cause of tensions within families; each family member wants to secure a large share of the 'land manna'.

5.1.2. Non-respect for traditional chieftaincies

The lack of respect for traditional chieftaincies is evident in the rejection of their decisions in disputes. These decisions are not always accepted and the protagonists very often resort to the police and judicial authorities. The Secretary General of Sapia says: '*From the moment the chief is not respected in the village, he may even summon someone from the village, and the man may not come here*'. Thus, the village chief is now seen as a simple inhabitant whose consultation is not necessary in case of conflict. Continuing, the Secretary General of Sapia adds: "*Our difficulties are that the protagonists do not understand us. They believe that when you draw a conclusion from a customary court, you are only against the one who is not right; often this is how they take. Otherwise on the ground, we want to make peace between the inhabitants of the village. We have nothing to gain, we have nothing to lose. We want peace, we want justice. Let the population understand us, the decisions we take are to make peace among family members.*

In integrated villages, a climate of тёйапсе prevails between the inhabitants and the village authorities, whom they accuse of complicity in most cases. As a result, the village authority struggles to gain acceptance for its desire to preserve social cohesion within the village.

5.1.3. The increase in land disputes in integrated villages

5.1.3.1.   The causes of land disputes

Land disputes are observed in Sapia, Tagoura and Zakoua with some particularities. Figure 17 shows the distribution of the causes of land disputes in each village.

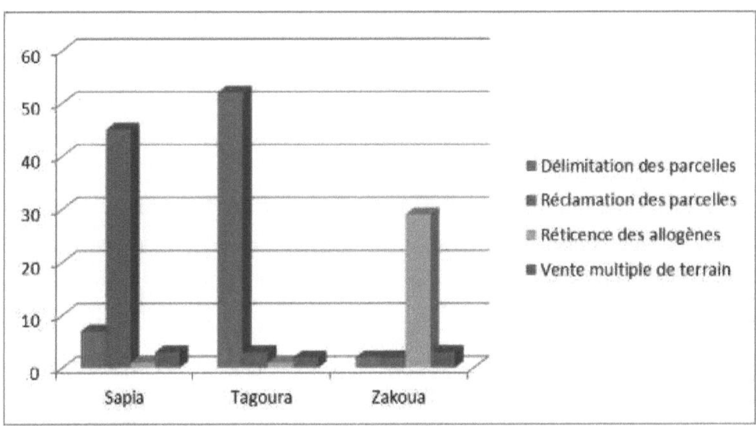

Source: Our survey, 2019

Figure 17: Distribution of causes of land disputes in integrated villages

The delimitation of plots (41%) and the reclamation of transferred plots (33%) are the main types of land disputes. This is followed by the reluctance of allogenes (21%) and the sale of the same land to several people (5%).

Figure 18 shows the distribution of types of land disputes in Sapia.

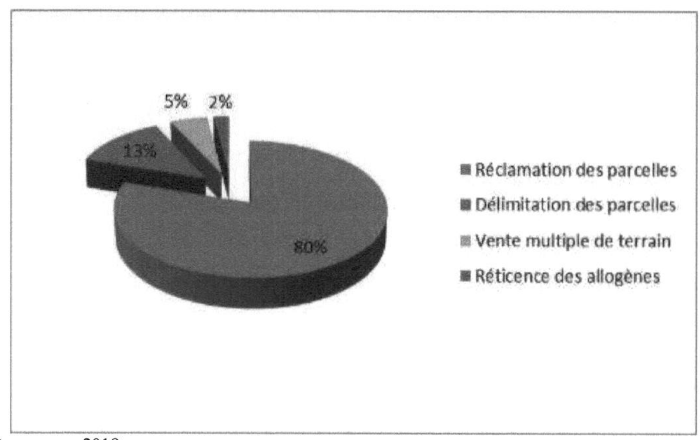

Source: Our survey, 2019

Figure 18: Breakdown of causes of land disputes in Sapia

In Sapia, land disputes Hёз to cёdёes plot claims (80%) are by far the most dominant since the dёbut of private allotments in 2014. Land disputes Hёз to plot dёlimitations (13%), multiple land sales (5%) and allogёnes reluctance (2%) together account for almost a quarter (20%) of these disputes. Indeed, the economic stakes involved in the sale of urban land have encouraged the questioning of land obtained from family members or relatives through family, friendly or neighbourly relations. The Secretary General of Sapia explains: "*He told me, little brother, you stay here, that's the limit between you and me. Today, I have done my food culture, and in the meantime, God calls him back. His children will say, our father worked here, so our allotment must take this part into account. I will say no, it is not normal, it is my elder brother who gave me the land for my food crops; or I gave him the land to grow his food crops. If not, that is not what you are saying, that is the limit between us. So at all times, these are the issues we are dealing with here. People are claiming too much of their parents' heritage.*

In Sapia, the reclamation of land patrimony, which used to be cёdёs, is almost permanent and has even become a commonplace societal fact, the settlement of which irritates the traditional chieftainship to the limit. The incessant claims to land patrimony create a situation of tension and mistrust within families. These claims also have as a corollary another type of land dispute Hё the sale of plots located at the boundaries of the parcels. Speaking about this type of dispute, the Secretary General of Sapia explains: "*You, he gave you your plot; your older brother also gave him; your younger brother also gave him. Everyone knows that he has his own limit. But if you overflow to enter Paul's house, Paul in turn will oppose the sale of the lots that are on the recovered boundary.*

In Sapia, the economic stakes raised by the subdivisions are at the origin of the multiplication of land disputes. Figure 19 shows the distribution of the causes of land disputes in Tagoura.

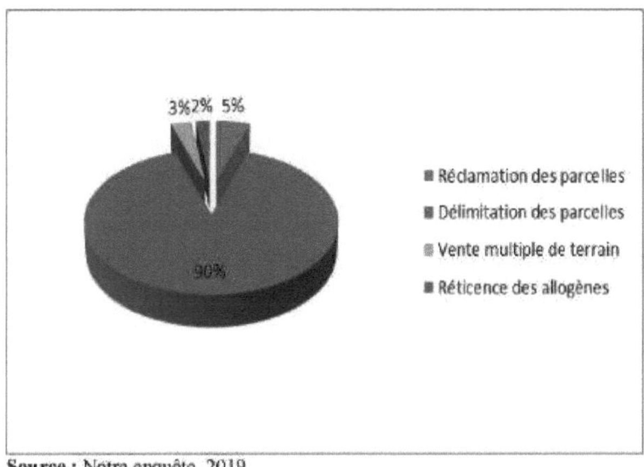

Figure 19: Distribution of causes of land disputes in Tagoura

In Tagoura, disputes over land are generally related to the delimitation of plots during subdivision operations (90% of the total), the reclamation of transferred plots (5%), the multiple sale of land (3%) and the reluctance of allogenes (2%). Disputes related to the delimitation of plots arise when, through carelessness or ignorance, one landowner encroaches on another's space. This type of dispute is usually settled amicably between the two neighbours or by sending an 'expert' to re-establish the boundaries of each party's plot. In view of the recurrent conflicts related to the delimitation of plots, it is now recommended that anyone initiating a subdivision should do so in the presence of their immediate neighbours.

Figure 20 shows the distribution of causes of land disputes in Zakoua.

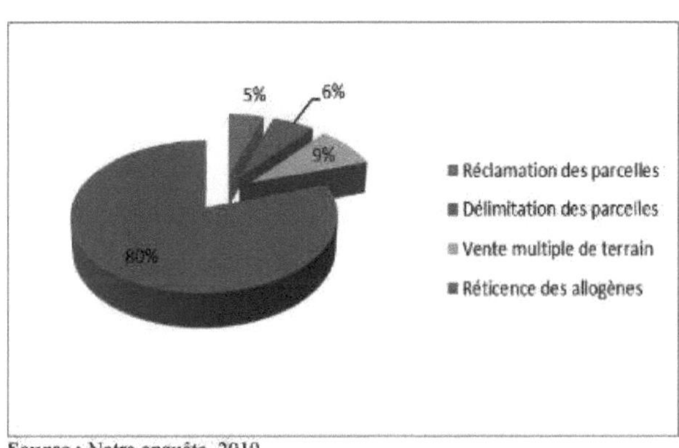

Figure 20: Distribution of causes of land disputes in Zakoua

In Zakoua, land disputes are primarily due to the reluctance of the allogenes (80% of the total) to subdivide the land on which they have been farming for decades. The importance of these disputes can be explained by the fact that the indigenous people of Zakoua have received large populations of non-indigenous people who have come to farm the forests. These allogenes have crëë vast plantations of cocoa and cafë. Since the city has reached these cëdëes plots, there is talk of dividing them up and making the lots available. This is when the landowner intervenes and asserts his right of ownership. Thus, conflicts often arise during the purging of cultivation rights or during the sharing of lots. This is reflected in the words of the chief of Zakoua: '*Our difficulties are that the allogenes are often reticent. It's the bush, you've bought, you haven't bought the land. You can't sell land. Now, from the moment you no longer have anything on that land, the land goes back to the landowner.*

Faced with the reluctance of allogenes who exploit the plots for agricultural purposes, a compromise is found in order to preserve an agreement between the landowner and the allogene. "Sometimes, *to avoid wrongdoing, we do equal shares. When you do an equal share, you are all happy, especially the non-native. And from then on, he becomes a full member of the family; when you have a problem, he comes to your rescue,*" explains the head of Zakoua.

The sharing of plots by consensus thus allows good relations to be maintained between the allogene and his guardian. The allogene always shows his gratitude to the landowner because he has enough to ensure his survival. He does not feel that an injustice has been created.

5.1.3.2.   Increasing land disputes

Land disputes have increased in integrated villages since the dëbut of private allotments in 2014. Figure 21 shows the evolution of land disputes handled by the chiefdoms over the last fifteen years.

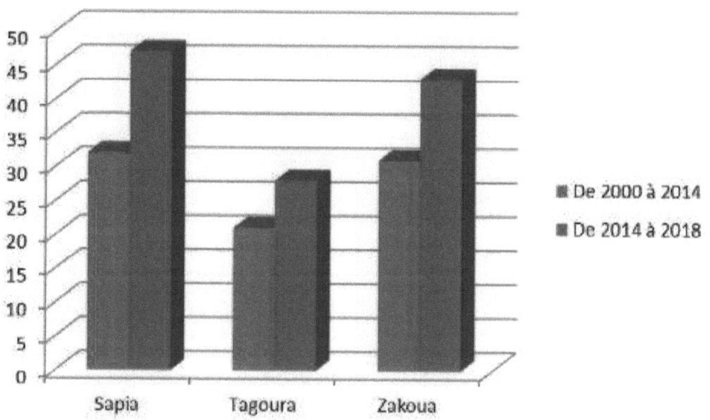

Figure 21: Evolution of land disputes in villages from 2000 to 2018

Analysis of the figure shows that land disputes have increased by 40% in the three villages, with 46% in Sapia, 33% in Tagoura and 39% in Zakoua respectively. In addition to land disputes related to the delimitation of plots, the reclamation of transferred plots and the reluctance of allogenes, which are specific to each village, land disputes related to the sale of a plot of land to several purchasers are transverse to the villages. Indeed, some customary owners sell a plot already allocated to one buyer to several other buyers. The new acquirers proceed with the transactions for the purchase of the land without prior verification. Disputes arise when one of the acquirers decides to develop the lot.

5.1.3.3.   The recommended framework for the settlement of land disputes

The choice of customary owners for the settlement of land disputes presents different patterns in the three villages. Figure 22 shows the distribution of recourse in the event of a dispute.

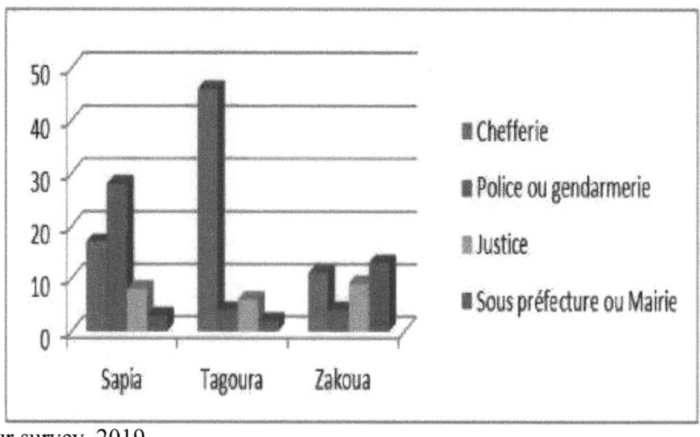

Source: Our survey, 2019
Figure 22: Distribution of dispute remedies in integrated villages

Overall, the chieftaincy (49%) and the police (24%) are the main recourse for customary owners in the event of a land dispute (73%). They are followed by recourse to the courts (15%) and to the sub-prefecture or town hall (11%). Figure 23 shows the distribution of recourse in the event of a land dispute in Sapia.

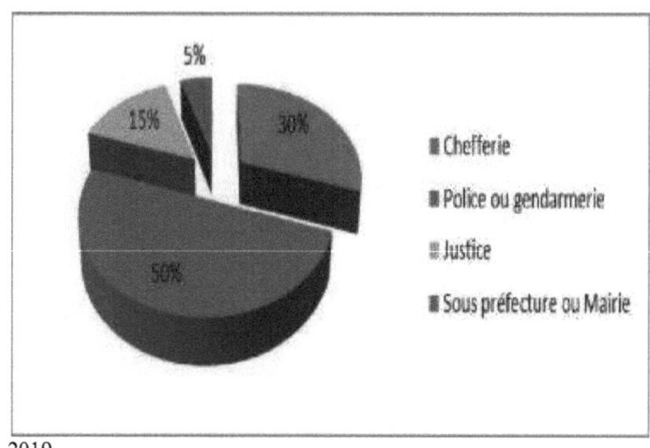

Source: Our survey, 2019
Figure 23: Distribution of remedies for land disputes in Sapia

In Sapia, the police are favoured for the settlement of disputes (50%). This choice is justified by the preponderance of young owners aged between 30 and 45 years (60%) with an average level of secondary education. This social category very often clashes with the ancient practices of land donation advocated by the chieftaincy. The customary landowners prefer the police or the gendarmerie for a settlement with modern principles.

In Tagoura, land disputes are settled by the chiefdom. Figure 24 shows the distribution of remedies for land disputes in Tagoura.

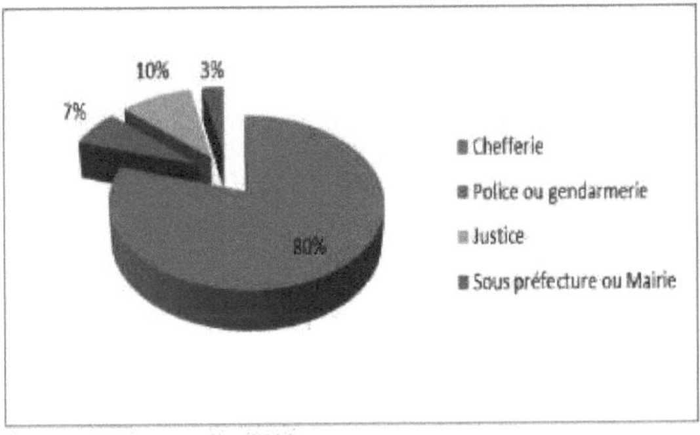

Figure 24: Distribution of remedies for land disputes in Tagoura

In Tagoura, the chiefdom is preferred in case of dispute (80%). This situation is explained by the organisational system of the village. The chiefs of the families, who are also the village notables, are responsible for settling land disputes within their families. The chiefdom is only called upon in the event of a major dispute that the head of the family has not been able to resolve. Figure 25 shows the distribution of remedies for land disputes in Zakoua.

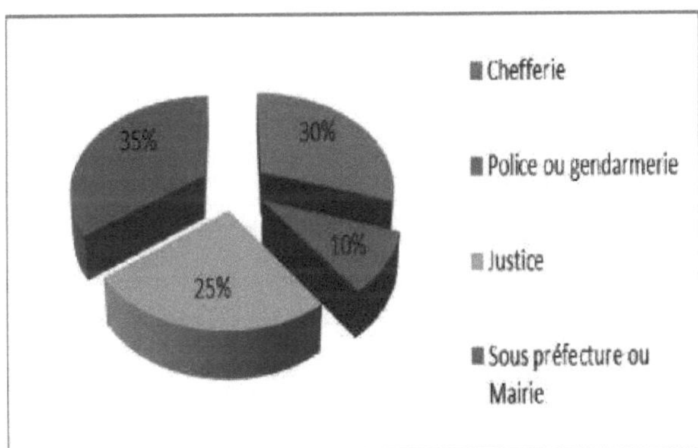

Source: Our survey, 2019
Figure 25: Distribution of remedies for land disputes in Zakoua

In Zakoua, the sub-prefecture is most often called upon in cases of land disputes (35%). Indeed, land disputes are generally between customary owners and allogenes who have come to exploit the land for agricultural purposes. Figure 26 shows the general trends in the choice of landowners for the settlement of land disputes.

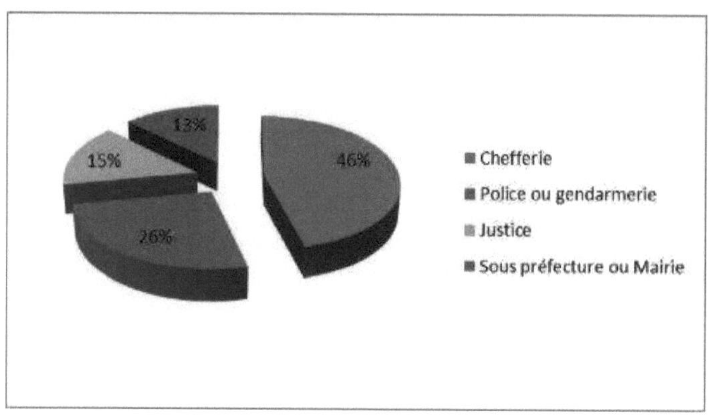

Source: Our survey, 2019

Figure 26: General trends in landowner choice for land dispute resolution

At the top of the list, the chiefs (46%) are more likely to be called upon in the event of a land dispute. Indeed, the village chiefs are retired civil servants who are well aware of the implications of the new land tenure provisions. They therefore combine elements of modern and traditional land law to separate the protagonists. When the dispute escalates, 26% of respondents prefer to go to the police or gendarmerie. Recourse to the courts (15%) is made for major disputes involving large areas of land, or for the sharing of a premium in the event of the purging of crop or land rights by the State.

46% of the respondents prefer to settle land disputes amicably under the supervision of the traditional chiefs. This view is well expressed in a popular saying by the chief of Zakoua: "*A bad settlement is better than a good trial. So if there is something going on, instead of going to the police, instead of going to court, it is better to sit down and find a solution.* The method of settling land disputes, which is done amicably, by consensus, based on compromise, therefore appears to be more judicious and the one recommended by the population because of the many advantages it offers. The chief of Zakoua sums up these advantages in the following terms: '*The courts can, for example, rule in favour of a third party, and then when you get to the village, if there is opposition, people will never work on the land, because you have to fear death. That is why it is better to sit down in the village and find a consensus. Even if there is a problem in the family, we will come and find a solution. And sometimes even the police, even the justice system, asks us to do this, because it is a way of helping them too.* This would facilitate the development of urban land and the integration of peri-urban villages into the city in a coherent way. Photo 2 shows the multi-purpose home in Tagoura.

Photo 2: The Tagoura multi-purpose home *(Clichi by the author, May 2019)*

The multi-purpose hall in Tagoura serves as a place for settling land disputes and other cultural activities in the village. It was built by the population with their own funds. This centre contributes to preserving the culture of the village in the face of rapid urbanisation.

5.1.4. The private housing estate: the end of land expropriation

Our survey reveals that 64% of Tagoura's landowners did not receive a compensation plot during the administrative subdivisions that took place in 1974. In fact, the town hall was opening up a subdivision frontage to satisfy the demand for urban land. The extension of the city was done on the surrounding rural land. The landowners were thus expropriated from their land that they used for agricultural purposes. In order to satisfy the dispossessed landowners, the developer-city council gave them plots as 'compensation'. In general, two lots per hectare were given to the landowner after the subdivision of his plot.

With private subdivisions, the landowners have the initiative to subdivide their plots and the bulk of the lots. The chief of Tagoura says: 'It *is better this way, instead of the state doing the parcelling out in the villages through the Mairie, without even compensating the de facto landowners most of the time.* From now on, the State recognises and consolidates customary rights, thus cancelling out the negative effects of administrative subdivisions. Indeed, in most cases, verbal agreements have been concluded between the municipal authority and the customary owner. The absence of a written document prevents any legal procedure to acquire a "promised" compensation plot. In the opinion of the landowners, private allotments naturally have many advantages. The chief of Tagoura says: *'The state has seen fit to give Cesar what belongs to Cesar. There are people who have been managing their land for thousands of years. Today, the State recognises that these people are the de facto owners, which is normal.* On the other hand, the difficulties encountered with private housing estates can be attributed to a gradual start and adaptation. *"With landowners it is much better. The problems we are experiencing, as it is a start, are perhaps due to the fact that we are still in the early stages. But as we go along, it will go well. Because today, if you go around the villages, all the village chiefs are practically people who used to work in the civil service and who are now retired. So they will know how to set up structures to better manage these problems, to better regulate and take precautions so that there are no disputes, so that there are no conflicts. Otherwise, the state has not done a bad job in asking the people to divide up their plots themselves,"* says the Tagoura chief. Land management will be better in a few years' time thanks to the supervision of the State through the Ministry of Construction and Urbanism and the structures set up in the villages.

5.2. The economic impact of urban dynamics on land management in the integrated villages of SAPIA, TAGOURA and ZAKOUA

5.2.1. Land speculation

Land speculation is taking place in integrated villages because of the market value of land. Land that was once given away and bequeathed is now a source of profit through the sale of land. There is a race by landowners to subdivide their plots, and an influx of surveyors and real estate agencies into the integrated villages, creating a profitable market for landowners. Very often, landowners cannot afford to pay the surveyors in cash, so they gamble on the offers of the surveyors and real estate agencies that court them. The sale of urban land is the main interest of the subdivision for the landowners. This reality is, however, subject to constraints as land is an inheritance that the head of the family shares with family members. The Secretary General of Sapia explains this reality in the following words: *'Lots, you don't sell everything. After the allotment, you have to pay the surveyor who did the work in kind, because you don't have any money in cash. So what you get is for your family. You reunite the family. The surveyor has taken his share, that is what is for the family. If there are many of you and the land is small, you are obliged to put yourselves together. The two of us, for example, can be given a plot, the other two can be given a plot, and so on. If the land is large and you are alone, you are obliged to sell part of the lots, and you develop the other lots for the future of your children"*. Figure 27 shows the distribution of landowners' choices for developing their land.

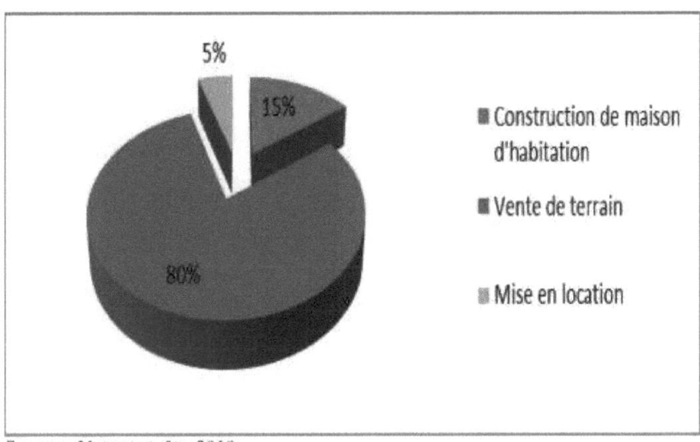

Figure 27: Distribution of landowners' choices for development of their plots

The analysis of Figure 27 shows that for the development of their lots, the sale of land is the preferred choice of the majority of landowners (80%), while 15% prefer to build dwellings. In fact, faced with speculation, landowners prefer to sell part of their lots in order to use the money earned to build houses for their families. The money collected from the sale of the lots is also used for funerals, social prestige or for the purchase of agricultural inputs. For peasants who do not have a stable income, 'selling a plot' is an opportunity to have money to meet their needs. Building a house on their plot is the choice of wealthy landowners (15%). A small proportion of landowners (5%) sign partnership contracts with real estate companies that build houses for them to rent. At the end of the period specified for depreciation, the houses revert to the owners of the developed plots. Unfortunately, it happens that the same plot is sold to several people. Figure 28 shows the factors of 'multiple land sales'.

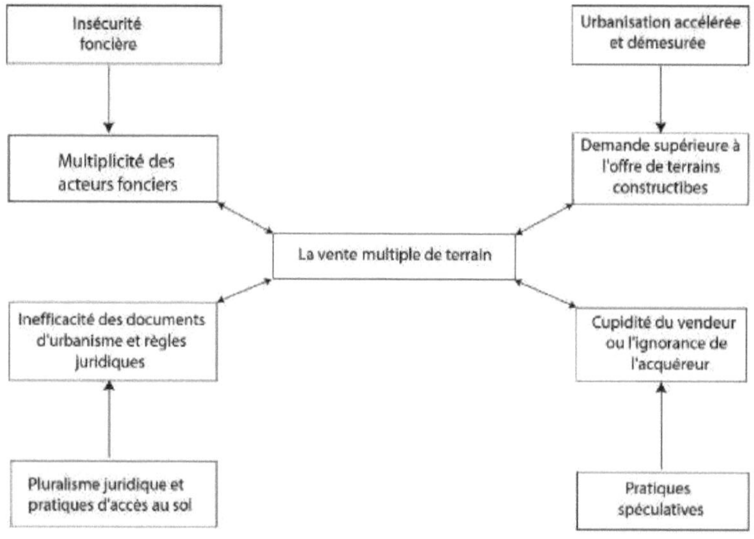

Figure 28: Factors in multiple land sales

The analysis of the graph reveals that the causes of 'multiple land sales' or the sale of a lot to several people are speculative practices, accelerated and disproportionate urbanisation, legal pluralism, land access practices and land insecurity. When speculative practices are at the origin of the multiple sale of land, it brings into play the greed of the seller or the ignorance of the acquirer. In Daloa, land speculation is the main cause of the 'multiple sale' of land.

5.2.2. Payment of property tax in integrated villages

Overall, landowners and land purchasers in the integrated villages of Sapia, Tagoura and Zakoua do not pay land tax despite administrative pressure. Negotiations are underway between the land registry and these village communities who consider the costs of the land tax high and arbitrary. *"How can we, our own houses that we have built, sometimes out of banco, sometimes out of bamboo, have people come to collect from us?*

The villagers refuse to pay for the urban land on which there are still basic mud houses that will be destroyed very soon. They also mention the fact that they have not been compensated during the administrative subdivisions. The chief of Tagoura expresses these claims, which are common to several villages in the commune, in the following terms: "There are *some villages in the commune that are still reluctant because they say that the town has developed to their detriment. They were given a lot of land and did not get any compensation. Now that they have been told that the land belongs to them, they can make their own plots, and then they are asked to pay taxes. What about the other allotments that have gone through where they have not been compensated? So how do they do it?* Due to the change in urban land legislation since 2014, landowners whose plots have been subject to administrative subdivisions feel entitled to reclaim their plots and refuse to pay the land tax. Figure 29 shows the distribution of landowners according to the payment of land tax.

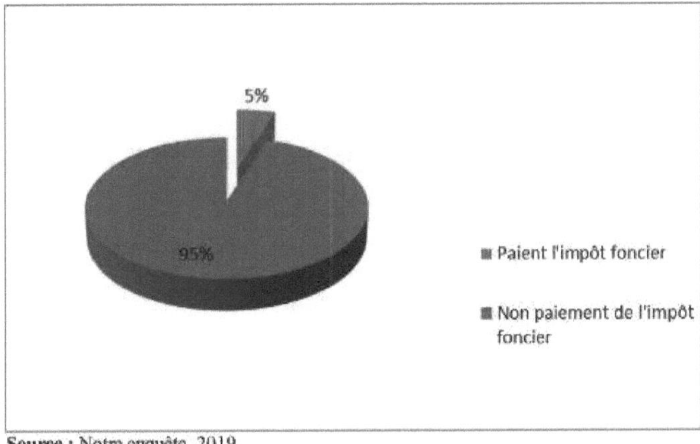

Source : Notre enquête, 2019

Figure 29: Distribution of landowners according to payment of land tax

The analysis in Figure 29 shows that landowners do not pay land tax (95%). The main reason ëvoiced by landowners who refuse to pay the land tax is that they are on their land. They also feel that they are not bërайacШ: retombës of the payment of the land tax. The reluctance of customary landowners to pay the land tax is also based on the lack of a legal document on the land and on the state of poverty due to the fall in the cost of agricultural raw materials.

The integrated villages are made aware of the payment of land tax by the land registry services due to their integration into the city. The collection of the land tax is carried out by the land registry service of the Directorate of Taxes. Everything is centralised in Abidjan before the theoretical transfer of 25% to the commune.

There are three ways to pay property tax:

- 1er case: For bare or insufficiently built-up land, the cost of the tax is obtained after multiplying a tax base corresponding to the current value of the land by the surface area. A commission sits in the municipality to fix the price per m^2 or the market value of the land by district according to the level of equipment. The state charges a rate of 1.5% of the sum thus obtained.

- 2eme case: The land is built and inhabited by the customary owner

In this case, a sum corresponding to a rate of 3% of the annual rent is fixed. The annual value of the rent if the house was rented out is determined and a rate of 3% is applied.

- 3eme cases: Built-up property assets with income

This catëgory concerns houses built and rented by the landowner. The cost of the land tax is йхё to 12% or 15% of the annual rent respectively when the house is rented by a natural person or a company. Of the 12% of the prelex tax, 3% goes to the municipality and 9% to the state. Table 15 shows the cost of the property tax paid by a landowner in Tagoura.

Table 15: Cost of land tax for a built-up area in Tagoura

Nature	Tax base	Rate (%)	Amount
Built-up property with income	1 260 000	9	113 400
	1 260 000	3	37 800
Total			151 200

Source: Our survey, 2019

Of the 151,200 F CFA to be paid for the annual land tax, an amount of 37,800 F CFA representing 3% of the total is paid into the communal fund. The amounts collected by the town hall for the land tax should allow the provision of infrastructure in the villages of the commune.

5.2.3. Economic conversion in integrated villages

In Sapia, Tagoura and Zakoua, because of the pressure on land due to the high demand for building land, the surrounding plots have all been subdivided and there has been a conversion of social categories. The Secretary General of Sapia explains that: *'the customary owners who have plots of land more than three kilometres away from the village still have reserves. While waiting for them to be allocated, they continue to live off their crops. Indeed, as the* city of Daloa expands, it is taking over rural land, the main resource used for agriculture, and its farmers are gradually turning to urban occupations. 53% of respondents are now engaged in a non-agricultural activity, either within the village (40%) or inside the city (60%) (Our survey, 2019). Increasingly, young people are reorganising to introduce urban activities into their lifestyle and are turning to masonry, plumbing and electricity. The women, on the other hand, are engaged in selling foodstuffs or in catering.

In Sapia, Tagoura and Zakoua, there are both formal and informal activities such as telephone booths, money transfer points commonly known as 'Orange money' or 'MTN money', petrol stations, car washes, hairdressing and tailoring salons, and outlets selling various items, etc. Figure 30 summarises the socio-economic issues of private housing estates in peri-urban villages.

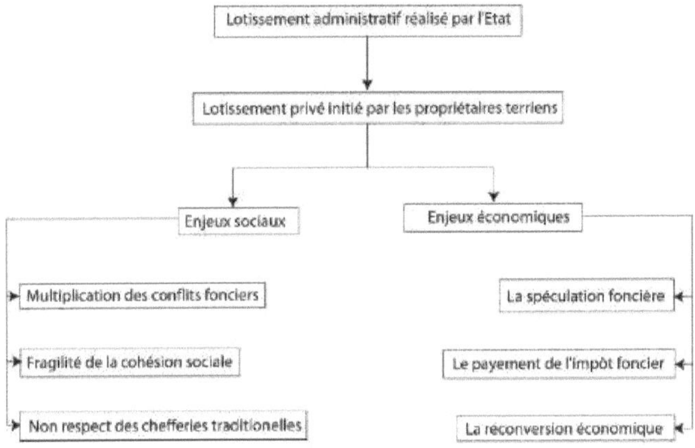

Source : notre enquête, Décembre 2018   Réalisation : N'GOTTA, 2018

Figure 30: Socio-economic issues of private housing developments

The shift from administrative subdivisions carried out by the state to private subdivisions initiated by landowners has led to a series of social and economic issues. At the social level, there has been an increase in land disputes, the fragility of social cohesion and the lack of respect for traditional chieftaincies. At the economic level, this change has led to land speculation, the payment of land taxes and the economic reconversion of villagers.

Conclusion

The sprawl of the city of Daloa has social impacts on land management in the integrated villages, including the fragility of social cohesion, the multiplication of land disputes and the non-respect of traditional chieftaincies. This integration also has economic impacts such as land speculation, payment of land taxes, economic reconversion of villagers etc.

ɸCHAPTER 6: THE SPATIO-ENVIRONMENTAL IMPACT OF DALOA URBAN DYNAMICS ON LAND MANAGEMENT IN INTEGRATED VILLAGES

Introduction

In Côte d'Ivoire, the control of urban sprawl is a major problem. This sprawl, which is achieved through the consumption of the land of the villages that are integrated, has multiple impacts that are observed more particularly at the spatial-environmental level. The interest of this chapter is to analyse the spatial-environmental impacts of the sprawl of the city of Daloa on land management in the integrated villages.

6.1. The spatial impact of urban dynamics on land management in the integrated villages of SAPIA, TAGOURA and ZAKOUA

6.1.1. Land use dynamics in integrated villages

The integration of Sapia, Tagoura and Zakoua into Daloa accelerates its spatial dynamics. Table 16 presents the distribution of the lots by village.

Table 16: Breakdown of lots completed by village

Villages	Allotments	Number of lots built
Sapia	56	1 204
Tagoura	58	1 008
Zakoua	36	621
Total	150	2 833

Source: Our survey, 2019

The housing developments in Sapia, Tagoura and Zakoua contribute to the spatial expansion of the city. Indeed, in its spatial growth, the city of Daloa consumes the land of the integrated villages, thus pushing the limits of the fields further and further away. The rate of expansion of Daloa is accentuated in its eastern and northern sectors towards Sapia and Tagoura respectively. In total, 2,833 lots are produced in the integrated villages of Sapia, Tagoura and Zakoua.

6.1.2. The preponderance of irregular housing developments in Sapia, Tagoura and Zakoua

Irregular subdivisions are predominant in integrated villages due to the cumbersome nature of the subdivision procedure. Indeed, the subdivision procedure appears to be long, complex and highly centralised, even though the initiative lies with the village communities and customary landowners. The landowner applies for a subdivision at the one-stop shop of the regional directorate of the Ministry of Construction and Urban Planning. The counter sends the application to the central directorate in Abidjan, which responds by requesting a commodo et incommodo investigation, with a view to determining whether the parcel to be subdivided is not subject to conflict. This enquiry is carried out by the town hall if the plot to be subdivided is within the municipal boundaries, or by the sub-prefecture if outside the municipal boundaries. Once this step has been completed, the gëomëtre proceeds to the polygon, which is the delimitation of the plot, and the town planner draws up the subdivision project, which consists of dividing the plot into lots according to the town's master plan. The subdivision project is in turn forwarded to the central directorate of the Ministry in Abidjan where only the Minister signs the subdivision approval order. The subdivision approval order marks the official existence of the lots on the plot. The subdivision file is handed over to the town planner who entrusts the gëomëtre with the demarcation. Figure 31 shows the distribution of approved allotments in the integrated villages.

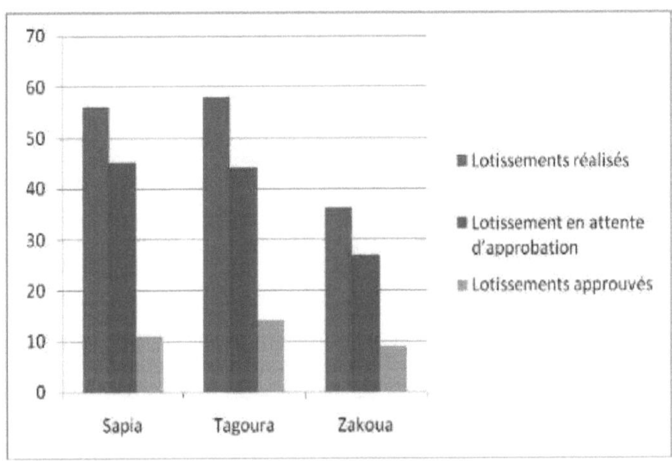

Source: Our survey, 2019

Figure 31: Distribution of approved housing developments in integrated villages

Analysis of the figure shows the prevalence of allotments awaiting approval in the three villages (77%). This is evident in each of the villages with 81% in Sapia, 76% in Tagoura and 74% in Zakoua. This situation is due to the cumbersome nature of the subdivision procedure. Indeed, all subdivision files are approved in Abidjan. This makes the procedure cumbersome.

The cumbersome subdivision procedure has two negative consequences: the production of subdivisions that are difficult to make viable and the development of an illegal land market. Landowners do not wait for the approval of the subdivision to proceed with the demarcation and sale of lots on the plot. The non-respect of urban planning norms due to the competition of actors and the bypassing of urban planners is at the origin of the production of allotments that are difficult to make viable. The Director of CIADEL explains that: *'the populations are authorised by the public administration to initiate subdivision operations. This measure certainly responds to a demand. However, it has been observed that the populations do not always rely on the administration or its local representatives. They engage in a subdivision operation. It is when they encounter difficulties (for example, during the sharing of lots) that they turn to the administration.*

In Daloa, out of the hundred or so subdivisions in the city, only forty-five are approved despite the regularisation of subdivisions applied by the Minister of Construction in 2015. A private individual said: *"The illegal land market exists because there is an irregular production of land products. In order to put an end to this illegal land market, it is the urban land production system that needs to be examined in order to know which aspects need to be changed or corrected*. The illegal land market is a major cause of the proliferation of slums within cities. The many volumes devoted to African cities in general, and those of the Ivory Coast in particular, show cities that are largely the result of subdivisions produced in circumvention of official rules. Of the artisans of these subdivisions, the customary owners are constantly recurring in the literature, regardless of the periods and fronts of evolution of the cities studied (BROU Emile, 2010). The customary owners, craftsmen and authors of the allotments were only able to engage in these operations of parcelling out and selling plots of land because of the existence of a market and the pressure of the demand for building land (YAPI-DIAHOU, 1991). The extreme centralisation of the urban land production system should be examined in order to clean up the environment and prevent the proliferation of precarious and illegal housing.

6.1.3. Under-equipping integrated villages with basic infrastructure

In Sapia, Tagoura and Zakoua, the populations are waiting for the construction of infrastructures planned on the administrative reserves left at the time of each housing estate. In fact, on each subdivision initiated by a landowner, there is an administrative reserve that should be used to provide the "hybrid district" with basic infrastructure (schools, clinics, markets, etc.). However, since the start of private housing developments in 2014, no infrastructure has been built on these administrative reserves in Sapia, Tagoura and Zakoua (Our survey, 2019). The Director of CIADEL notes that: *'the cities of Cote d'Ivoire are not equipped with infrastructure in time. The provision of infrastructure (roads and other networks, water, electricity and social infrastructure) does not follow urbanisation. Often, the infrastructure is put in place long after the population has settled.* Despite the housing developments, Sapia, Tagoura and Zakoua are not serviced and there is a lack of drinking water supply, the absence of roads and poor electrification, apart from the streetlights that line the national roads that cross them.

Moreover, Sapia, Tagoura and Zakoua are under-equipped in terms of basic modern infrastructures in the face of a growing resident population due to residential mobility towards the periurban area, confirming STEINBERG J. (2003, p. 82): "In the periurban area, one almost always finds a more or less accentuated delay in relation to the needs: this is the famous "vicious circle" of periurbanisation, which often leads to a lack of access to basic services. 82): *"In the periurban area, we almost always observe a more or less accentuated delay in relation to the needs: it is the famous "vicious circle" of periurbanisation, which often leads to the "flight ahead" of the territorial authorities: the demand for equipment induces an increase in the population, which in turn stimulates a new demand for equipment and so on. Despite the great efforts made, the peri-urban activities and facilities are not sufficient to meet the local demand for jobs or services. These areas are characterised by the constant need to have recourse to the main urban agglomeration, which leads to considerable mobility of the peri-urban population, whether for work, shopping, study or leisure".*

60% of the respondents visit the city centre of Daloa on a daily basis, where the commercial district is unavoidable. "It is the place where all the administrative functions that are useful to them are located: the town hall, the prefecture, the tax administration, ... and the majority of banking establishments for matters of prestige" (GOHOUROU F. et al., 2017).

The integrated villages of Sapia, Tagoura and Zakoua are under-equipped with modern infrastructure: dispensaries, primary schools, modern markets, etc. Only one school complex with three schools and a health centre has been built in Sapia. Tagoura has a dispensary, a primary school and a water pump. Zakoua has a primary school and a modern market under construction. Part of Tagoura's land is being used to extend the Jean Lorougnon Guëdë Umversite in Daloa. Photographs 3 and 4 show public infrastructures built in Sapia and Tagoura.

Plate 1: Public infrastructure built in Sapia and Tagoura

Photo 3: School group of three schools in Sapia Photo 4: Dispensary built in Tagoura
(Author's copy, May 2019)(Author's copy, May 2019)

6.2. The environmental impact of urban dynamics on land management in the integrated villages of SAPIA, TAGOURA and ZAKOUA

6.2.1. Proliferation of unbuilt lots

In Sapia, Tagoura and Zakoua, we note the abundance of unfinished or insufficiently built lots in the form of unfinished houses, bare land, or interstitial spaces used for intra-urban agriculture. These lots are usually disputed land for which no solution has yet been found, or land owned by landowners unable to develop it. The proliferation of undeveloped plots of land makes the periurban landscape look unfinished and untidy. This situation, which is a consequence of the competition of the actors in land production, forces new land buyers to settle on isolated plots of land, sometimes in the middle of the forest, in the hope of being caught up by the urban construction a few years later. Housing estates with a better level of facilities contain plots that have been desperately empty for several years. And as nature abhors a vacuum, these plots are naturally transformed into cultivation space (N'GUESSAN V. 2003, p. 5). Bare land and unfinished houses are sometimes used for agricultural purposes, reflecting the competition between urban advance and agricultural production space. Figure 32 shows the distribution of built-up areas by village.

Source: Our survey, 2019

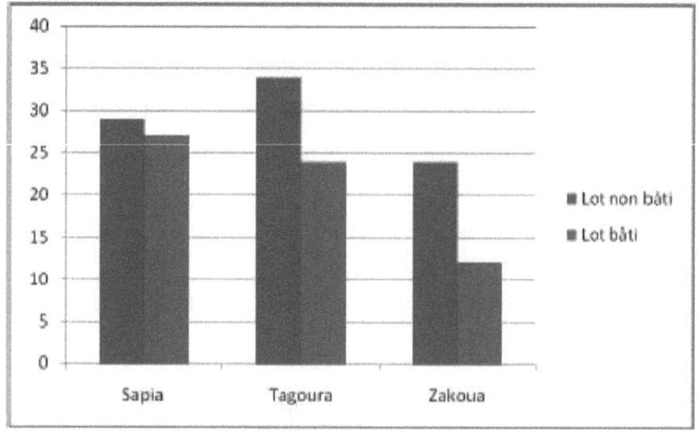

Figure 32: Distribution of built-up areas by village

Analysis of Figure 32 shows that the rate of development of the plots is 48% in Sapia, 41% in Tagoura and 32% in Zakoua, i.e. an average of 42% for all three villages surveyed. These rates show that the habitat is densified in Sapia and Tagoura. Table 17 shows the reasons for the lack of development of the lots by village.

Table 17: Distribution of reasons for insufficient development of plots by village

Villages	Lack of financial means	Land dispute	No land title
Sapia	15	38	3
Tagoura	42	12	4
Zakoua	10	23	3

Source: Our survey, 2019

The analysis of the table shows that land disputes (49%) and lack of financial means (45%) are the main reasons for insufficient development of the lots.

In Sapia, land disputes (68%) are the major cause of insufficient development of the lots. Indeed, the construction of houses is stopped when disputes arise over land. There is also a land dispute between Sapia and the neighbouring village, Gogoguhe. The two villages are in conflict over the boundaries of their land holdings and over the sale of lots in certain areas.

In Tagoura, the lack of financial means is the main reason for the insufficient development of the plots.

In Zakoua, land disputes (63%) are ëvoquës to justify the insufficient development of the lots. In fact, many parcels of land that have been parcelled out are sources of dispute in Zakoua. The landowners believe that the quotas of compensation lots set during the administrative subdivisions have not been reached for certain plots. As a result, they demand the eviction and destruction of all constructions on these plots. The purchasers having built on these plots have received letters of summons demanding the eviction. While waiting for the court's decision and the resolution of these disputes, large built-up areas are partially or not inhabited. The empty spaces between the buildings are used for agricultural purposes (photos 5 and 6).

Plate 2: Empty spaces used for agricultural purposes in Zakoua

**Photo 5: Maize crops on a field Photo 6: Banana crops on a field
unfinished house in Zakouanu   in Zakoua.**
(Author's copy, May 2019)(Author's copy, May 2019)

These photographs show mai's (photo 5) and banana (photo 6) being grown on the space of an unfinished house and a bare plot of land in Zakoua respectively. For their part, the owners of these plots claim that they do not have the means to proceed with construction. Until they have the necessary means to build houses, the owners prefer to use these plots for agricultural crops rather than leaving them uncultivated.

6.2.2. Scrub that dots the city

The scrubland that occupies bare land constrains urbanisation in increasingly remote areas. In Sapia, Tagoura and Zakoua, the landscape reveals a living environment where the bushes are spread between the houses, sometimes isolating them. The presence of scrub suggests poor land management in which the subdivisions made do not always correspond to town planning standards because this scrub sometimes occupies land not taken into account between two housing estates. This almost permanent undergrowth can be seen on undeveloped or insufficiently developed land, along the asphalt roads that cross these villages and along the alleys. Photograph 7 shows the scrub that isolates the houses in Tagoura.

Photo 7: Scrub isolating dwelling houses in Tagoura (*Author's photo, May 2019*)

Photo 7 shows that in the intëgrës villages (Sapia, Tagoura and Zakoua), the rate of land occupation by brush (40%) is close to that of the dwelling houses. This favours the diffuse extension of the city in this sector. This alarming situation of overgrowth that is flourishing in these urbanised areas does not seem to concern the municipal authorities, whose actions are not evident. The Director of CIADEL states that "*the management of the environment by the municipal services is mediocre. It needs to be improved. This management has several dimensions (administrative, technical, financial). For better management of the environment by the municipal services, there must first be a good understanding of the issue by the actors at the municipal level. Then, a good strategy must be put in place.*"

The spatio-environmental challenges of integrating Sapia, Tagoura and Zakoua in Daloa are summarised in Figure 33.

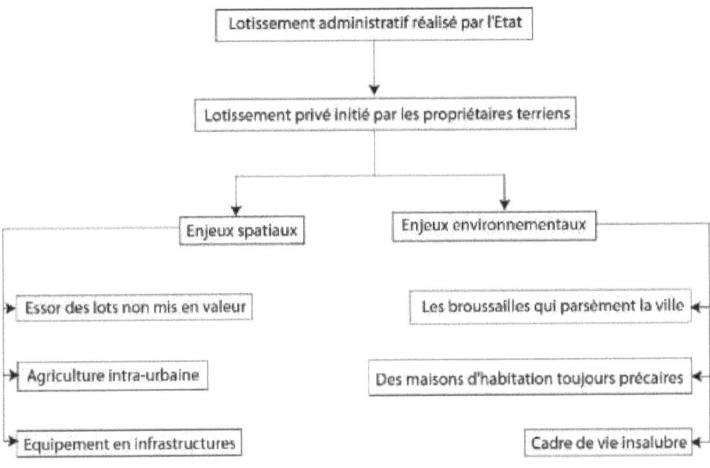

Figure 33: Spatio-environmental issues of private housing estates

Private housing developments have spatial and environmental implications. On the spatial level, the proliferation of undeveloped plots, the development of intra-urban agriculture and the under-equipment of basic infrastructure in peri-urban areas are observed. At the environmental level, the proliferation of undeveloped plots, the permanence of the bushes that dot the city, etc., can be observed.

Conclusion

The integration of the villages of Zakoua, Tagoura and Sapia into the city of Daloa has an impact on land management, which manifests itself at the spatial level through the sale of large areas of land that contribute to the rapid expansion of the city, the development of intra-urban agriculture, etc. The environmental impacts of this integration are the proliferation of unbuilt plots and unfinished houses, the brush that scatters the city, etc. The environmental impacts of this integration are the proliferation of unbuilt lots and unfinished houses, the bushes that dot the city, etc.

CONCLUSION OF PART THREE

The integration of the villages Sapia, Tagoura and Zakoua into the urban space of Daloa has socio-economic and spatial-environmental impacts on land management. At the socio-economic level, we observe the fragility of social cohesion, the non-respect of traditional chieftaincies, land speculation, the economic reconversion of villagers, etc. At the spatial-environmental level, we have the proliferation of undeveloped lots, the poor infrastructure equipment, the insalubriousness of peri-urban spaces, etc.

Consequently, hypothesis 3 of the study, which states that: "land speculation is at the origin of land disputes and the rise of undeveloped plots of land that are marring the urban landscape of Daloa", is confirmed.

In Côte d'Ivoire, the control of urbanisation aтоrcěе since the 1950s is a major probiem. The spatial extension of existing cities and the creation of new cities and/or galloping urbanisation are the manifestations of this phenomenon. Urban growth is achieved through the integration of peripheral villages and the consumption of their land holdings. Since 2014, it is the village communities and the holders of customary land rights who are proceeding with the subdivision of their plots.

In the first part of this study on "Urban dynamics and land management in villages integrated into the city: the case of Sapia, Tagoura and Zakoua in Daloa (Cote d'Ivoire)", our results showed that the demographic and spatial dynamics of Daloa are remarkable. Due to the rapid expansion of the urban space, the villages of Balouzon, Bribouo, Gbokora, Zakoua, Tagoura and Sapia, etc., which were previously located on the margins of the city, are now part of the city. This spatial growth has taken place through several restructurings of spontaneous neighbourhoods. Thus, hypothesis 1 of the study, which states that: "In view of the remarkable spatial dynamics of Daloa, the villages of Sapia, Tagoura and Zakoua are now integrated into the city" is confirmed.

In the second part, the results showed that village communities and customary land rights holders in Zakoua, Tagoura and Sapia are subdividing their plots through surveyors and real estate agencies. This allows landowners to profit from their land holdings through negotiations. Consequently, hypothesis 2 of the study, according to which 'land production in the integrated villages of Sapia, Tagoura and Zakoua is carried out through internal organisation and subdivision operations', is confirmed.

The third part of the study highlights the socio-economic and spatio-environmental impacts of the urban dynamics of Daloa on land management in the integrated villages. Land has acquired a market value. The economic stakes that appear due to land speculation cause land disputes that undermine social cohesion, lead to an increase in undeveloped land and the development of intra-urban agriculture etc. The landscape of the city is unfinished with the presence of bushes in some places, the absence of roads and various networks (V.R.D) and infrastructure equipment. Therefore, hypothesis 3 stating that "Land speculation is at the origin of land disputes and the rise of undeveloped plots of land which are ugly for the urban landscape of Daloa" is confirmed.

In conclusion, we note that Daloa's expansion is rapid. Today, the city has reached the villages of Sapia, Tagoura and Zakoua, which were initially located in the periphery. This spatial growth has a socio-economic and spatio-environmental impact on land management in the integrated villages. Thus, the hypothesis дёпёгаlе of the study: "The spatial dynamics of the city on its margins has a socio-economic and spatio-environmental impact on land management in the integrated villages Sapia, Tagoura and Zakoua" is confirmed.

In order to further develop this study, the following research areas can be explored:
- La probkmatique de l'accёз au foncier a Daloa ;
- Study of basic facilities and infrastructure in urban villages integrated with the city.

Far from being a completed study, the present contribution opens up lines of research and perspectives of analysis that can be explored.

BIBLIOGRAPHIC REFERENCES

ADOU (A. Giscard), YAO (K. E), GOUAMENE (D. C), (2017), *"Competition for the occupation of lowlands in urban and peri-urban spaces in Daloa: between food production and real estate promotion"*, African Journal of Scientific and Technological Communication, Sërie Sciences Sociales et Humaines, IPNETP, Cote d'Ivoire, N° 49, May, ISSN: 2-909426-32-7, EAN: 9782909426327, pp. 6455 - 6476.

European Environment Agency (2006), *Urban Sprawl in Europe*, 114 p.

AKPINFA (D. E), 2006, *Problematique de la gestion fonciere dans les centres urbains secondaires du Benin:cas de Glazoue et Dassa-Zoume*, mëmoire de maitrise, Universite d'Abomey-Calavi, Benin, 76 p.

ALLA (D. A), 1991, *Dynamique de l'espace periurbain de Daloa*, these de Doctorat, IGT, Abidjan 453 p.

ANGUELETOU A. (2007), *Etalement urbain et periurbanisation des grandes metropoles indiennes, le cas de Mumbai*, Presses universitaires d'Orleans, pp. 55-74.

ATLAS de la population et des equipements de la Cote d'Ivoire, 2007, p 9.

ATTA K. (1975), *Etude des espaces urbains des villes de Cote d'Ivoire par l'interprétation de laphotographie aerienne*, Master's thesis, Abidjan IGT, 120 p.

ATTA K. (1978), *Dynamique de l'occupation de l'espace urbain etperiurbain de Bouake (Cote d'Ivoire)*, Thëse de doctorat, EHSS Paris, 309 p.

ATTA K. (1984), 'Urbanisation et speculation fonciere a Bouake', in *Annales de l'Universite d'Abidjan*, serie G (geographie), tome XIII : pp. 5-51.

BAILLY A., FERRAS R. (2004), *Elements d'epistemologie de la geographie*, Paris Armand Colin, 190 p.

BEAUCIRE F. et al (2007), *Etalement urbain et action publique : l'exemple de la Seine- et-Marne*, Gwenn PULLIAT, 107 p.

BEAUJEU-GARNIER J. (2006), *geographie urbaine*, Armand Collin, Paris, 349 p.

COTTEN A-M. (1974), *Un aspect de l'urbanisation en Cote-d'Ivoire*. In: Cahiers d'outre-mer. N° 106 - 27e annee, pp. 183-193.

CUNHA A. et al (2000), *Structures et dynamiques socio-demographiques des espaces urbains*, Observatoire de la ville et du developpement durable, Institut de

дёодгарЫе - Faculte des Gëosciences et de 1 Environnement, Universite de Lausanne, 103 p.

DAVODEAU H. (2005), *Les paysages, une nouvelle preoccupation dans la gestion des espacesperiurbains,* Cahiers d'economie et sociologie rurales, n° 77, 20 p.

DESJARDINS X. (2017), "*les espaces periurbains : une marge urbaine a soigner ou une nouvelle banalite territoriale ?*", Bulletin de 1 Association de gëographes frangais, 14p.

DJEKI J. (2003), *politique urbaine et dynamique spatiale au Gabon : le cas de Port - Gentil,* Thëse de Doctorat, Universe Laval, Quebec, 402 p.

DUPONT V. (2005), *peri-urban dynamics : population, habitat and environnement on the periphery of the large indian metroplises,* Arcview of concepts and general issues, CHS, Occasional paper n°14, pp. 3-19.

EGGERICKX et al (2002), *Demographie et developpement durable. Migrations et fractures socio-demographiques en Wallonie* (1990-2000), Liëge, Louvain-la- Neuve, SSTC, 208 p.

ENAULT C. (2003), *Vitesse, accessibilite et et etalement urbain : analyse et application a l'aire urbaine dijonnaise.* Laboratoire de Gëographie THEMA - CNRS, UFR Sciences Humaines, Dëpartement de Gëographie, Universe de Bourgogne, 69 p.

FOFANA B. (2015), *Dynamique urbaine et problemes environnementaux a Bouake*, mëmoire de Master, Universe Alassane Ouattara, Cote d'Ivoire, 187 p.

FUJITA, M. and THISSE, J.F. (2003), *Economie des villes et de la localisation*, collection " Economie, sociëtë, region ". Brussels, 88 pp.

GNABELI R. (2008), *La production d'une identité autochtone en Cote d'Ivoire,* Journal des anthropologues, p. 1-19.

GNABELI R. and LOGNON J. (2011), *Urban pressure and the identity of relict villages in Ivorian cities,* KASA BYA KASA, n° 19, pp. 20-33.

GOHOUROU F et al. (2017), *Mobilite et pratiques socio-spatiales des populations des zonesperiurbaines de Daloa,* journal scientifique europëen, vol 13, 16 p.

GUMUCHIAN H., MAROIS C. and FEVRE V., 2001. *Initiation a la recherche en geographie: Amenagement,* dëveloppement territorial, environnement, Paris, 2001, 425 p.

HAERINGER P. (1969), *Structure fonciere et création urbaine a Abidjan,* cahier d'etude Africaine, n°34, pp. 219-270.

HAERINGER P. (1970), *La dynamique de l'espace urbain en Afrique noire et a Madagascar: problemes de politique urbaine,* Colloque, 15 p.

HAERINGER P. (1979), *Occupation de l'espace urbain et periurbain, Atlas de Cote d'Ivoire,* Abidjan, Ministere du plan, Universite d'Abidjan, ORSTOM, 58 p.

JAUZE J-M. and NINON J. (1999), *Dynamiques et expressions de la periurbanisation a la*

Reunion. In: Cahiers d'outre-mer. N°206 - 52e annee, Avril-juin 1999. pp. 143-168;

JULIEN P. (2005), *Analyse critique de la pertinence de l'aire urbaine pour etudier l'etalement urbain,* Groupe Etudes Periurbain & Programme IUD 8, 30 p.

KADET (G. B), 1999, *Dynamique spatiale et gestion municipal de Guiglo, dans l'Ouest ivoirien,* thèse de Doctorat troisieme cycle, Paris universite 369 p.

KOFFI (E. B), 2010, "*Les lotissements irreguliers et la production de la ville : les quartiers Ayakro et Sagbe a Abidjan*" in les cahiers d'Afrique du CERASA, Paris, 13 p.

KOUAKOU B. et al (2015), *les conflits fonciers dans la ville de Korhogo*, Revue Africaine d'Anthropologie, Nyansa-Po, n° 19, 20 p.

KOUAME G. et al (2016), *Cadre d'analyse de la gouvernance fonciere de la cote d'ivoire*, World Bank Group, 186 p.

KOUASSI K. (2012), *Insalubrite, gestion des déchets menagers et risque sanitaire Infanto-Juvenile a Adjame,* IGT, 597 p.

KOUKOUGNON W.G., 2012, *Milieu urbain et acces a l'eau potable : cas de Daloa (centre-ouest de la cote d'ivoire)*, These uniques de Doctorat, UFHB, Abidjan- Cocody, IGT, 371 p.

LASSERRE G. (1970), *La dynamique de l'espace urbain a Libreville : reglementation fonciere et morphologie des quartiers*, CNRS, Paris, 23 p.

MCU-DCU, "*Plan d'urbanisme de reference, programme d'action quinquennal 19811985*", BCEOM, 1981, p. 79.

MEMEL (F. A), 2006, *Dynamisme urbain et gestion fonciere a Grand Bassam,* Master's thesis, IGT, Abidjan, 140 p.

MUGANZA (2009), *l'impact de l'operation de conversion des titres immobiliers sur les conflits fonciers en RDC*, Universite Lubumbashi-Licence en Droit, 53 p.

N'GUESSAN V. (2003), *processus d'extension spatiale urbaine et subsistance des activites agricoles a Bouake (avant le 19 Septembre 2002), 9* p.

United Nations, 1995. Population and Development, vol. 1: Programme of Action adopted at the International Conference on Population and Development, 89 p.

NGANA F. (2004), *Representation des espaces urbains et processus migratoire des populations marginisees en Centrafrique*, Thëse de geographie, Universite Denis Diderot-Paris 7, Paris, France, 438 p.

NOUGAREDES B. (2011), *Quelles solutions spatiales pour integrer l'agriculture dans la ville durable ? Le cas des " hameaux agricoles " dans l'Herault*, Presses universitaires de Rennes, Norois, n° 221, 2011/4, p. 53-66.

UN-HABIAT (2011), *Third African Ministerial Conference on Housing and Urban*

Development, 22 - 24 November, BAMAKO MALI, 30 p.

PLATEAU J.P., (1998), *Une analyse des theories evolutionnistes des droits sur la terre. In Quelles politiques foncieres pour l'Afrique rurale? Reconciling practices, legitimacy and legality*, Delville P.L., Paris, Karthala, pp. 120-129.

POLESE M. (2010), *The role of cities in economic development: another look,* Working paper, n° 2010-4, 37 p.

PRAT A. (2010), *Ouagadougou, capitale sahelienne : croissance urbaine et enjeux fonciers*, mappemonde, 24 p.

SAID B. (2008), *La ville en question - analyse des dynamiques urbaines en Algerie.* Penser la ville - approches comparatives, Khenchela, Algerie,14 p.

SCHEIBLING J., 2004, *Qu'est-ce que la geographie*, Paris, 197 p.

SERRANO J. (2007), *les espaces peripheriques urbains et le developpement durable : analyse d partir du cas de l'agglomeration tourangelle*, Universite de Tours, Cites, Territoires, Environnement et Societes CNRS, 43 p.

Statistical Services of the Statistics Division, Department of Economic and Social Affairs of the United Nations Secretariat, New York, May 2018, 283 pp.

STEINBERG J. (2003), *La periurbanisation en France (1998-2002)*, Geolnova 7, 12 p.

VANIER M. (2000), "Qu'est-ce que le tiers espace? Territorialites complexes et construction politique", *Revue de Geographie Alpine*, vol. 88, n° 1, pp. 105-113.

VERON J. (2008), *Enjeux economiques, sociaux et environnementaux de l'urbanisation du monde,* Paris, La Decouverte, p. 132.

VILLA M. (1996), *Distribution espacial y migration de la poblacion de America latina,* in Coordination Dora E. Celton, Migration, regional integration and productive transformation, Centro de Estudios Avanzados, Universidad Nacional de Cordoba, pp. 9-87.

YAO (K. E), 2014: *L'impact des unites industrielles de transformation du bois sur le developpement urbain a Daloa*, unique PhD thesis in geographie, Universite Felix I louplioue't-Boigny d'Abidjan-Cocody, IGT, 291 p.

YAO E. et al (2017), *Urban dynamics and indigenous land management strategies in Zoukougbeu (RCI)*, Longbowu, n°004, pp 491- 513.

YAO, (N.B), 2009: *La Maitrise du foncier dans le Processus du Programme special de Transfert de la Capitale a Yamoussoukro*, ISCAE, 65 p.

YAPI (D. A), 1981, *Etude de l'urbanisation de la peripherie d'Abidjan : l'urbanisation de*

Yopougon, these de Doctorat, 322 p.

YAPI DIAHOU Alphonse, 1991, *notes sur les demandeurs de terrains a batir a Daloa*, ORSTOM, Fond documentaire, p22.

YAPI-DIAHOU A., (1991), *Les detenteurs coutumiers, les citadins et l'Etat dans la course pour l'acces au sol urbain a Abidjan*, In E. Le Bris et al, Contribution a la connaissance d'un droit foncier intermediaire dans les villes d'Afrique de l'Ouest, ORSTOM, IFU, IRSSH, pp. 11-84.

YOUANA J. (1996), *Gestion fonciere et discipline urbanistique au Cameroun : Apports et limites du permis de construire*, revue de geographie du Cameroun, 41 p.

ZIAVOULA R., (1987), *La course a l'espace urbain : les conflits fonciers a Brazzaville* CERPAD, Universite Marien-Ngouabi, 29 p.

1. INTERVIEW GUIDES

1.1. INTERVIEW WITH TRADITIONAL CHIEFS

Date and place of the interview

1.1.1 Identification

Sex ... Age ... Ethnicity/Nationality Occupation Level of education
What value do you place on land
What is the role of the chiefdom in land management in Daloa
What are the conditions of access to land
What are the required documents?
What is the process of access to land
What are the sizes and prices of the plots
Is there a landowners' association in the village?   YesNo
If so, how does it work? How does it work?
What are his areas of competence?
What land conflicts have you had to deal with? In which sectors?
Are land conflicts related to :
a- Delimitation of plots / b Sale of land / c Subdivision

What causes conflict?
How (by what) do you recognise the boundaries of the plots
How are land disputes resolved?
What is given as compensation
What are the consëquences of land conflicts
Identification of purchasers

Sex . Age. Ethnicity/Nationality Profession . Level of education .

1.2. INTERVIEW WITH THE MAYOR OF DALOA

Date and place of interview

1.2.1 When was Daloa erected as a (fully functioning) commune?
1.2.2 Identification of the actors (involved) in land management in Daloa?
1.2.3 What is the role of the town hall in land management?

1.2.4 Is there a master plan for the city?
1.2.5 Are there any plans for the details of the Zakoua, Tagoura and Sapia housing estates?

1.2.6 What housing developments have been initiated by the town council since its creation to date?

1.2.7 When was the last town hall development?
1.2.8 Who is responsible for the sale of urban land at the level of the town hall?

1.2.9 What are the sizes and prices of the lots (plots)?

1.2.10. Does the town hall issue certificates?
1.2.11 Acquirer identification Gender...Age...

Ethnicity/NationalityOccupationEducation level

1.2.12. What is the nature of land conflicts?

1.2.13 How are land conflicts managed?

1.2.14 What do the lot applicants blame the town hall for or what complaints are registered?

1.2.15 What are the means of redress in case of dissatisfaction?

1.2.16 What are the difficulties encountered by the town hall in land management?

1.2.17. What are the prerogatives of the town hall in land management?

1.2.18. What is your opinion on indigenous land management (since 2014)?

1.2.19. What is the status of the collaboration with the Ministry of Construction?

1.2.20. What is the status of collaboration with customary owners?

1.3. INTERVIEW WITH THE HEAD OF THE TECHNICAL DEPARTMENT OF THE TOWN HALL

Date and place of interview

1.3.1. The housing estate

1.3.1.1- What is the procedure of the allotment in Daloa?

1.3.1.2- What are the reasons for opening a new housing estate frontage?

1.3.1.3- Who decides on the opening of a housing estate frontage? When do they decide?

1.3.1.4- What is the revolution in housing developments in Daloa?

1.3.1.5- How many lots do you have for each development?

1.3.1.6- What is the size of the urban perimeter? Before 2014 / After 2014
1.3.1.7- How many lots are there in the city in total?
1.3.1.8- Are they all from a regular housing estate?
1.3.1.9- When was the last development?
1.3.1.10-How many lots were distributed?
1.3.1.11-Which sectors are concerned?

1.3.1.12-Is it still possible to open up other housing fronts? In which geographical area of the city?

1.3.2. Allocation of urban lots

1.3.2.1- Are the lots regularly requested?
1.3.2.2- How many batch requests do you know of that are still pending?

1.3.2.3- What are the criteria for electing beneficiaries?

1.3.2.4- Is there a category of people who demand more than others?
Which ones?
1.3.2.5- What are the social ranks of the lot applicants?
1.3.2.6- Do you conduct a baseline survey on willingness to build?
1.3.2.7- How much does an urban plot cost?
1.3.2.8- Are there cases of lots being allocated to several people?
1.3.2.9- What is the waiting time between application and allocation of a lot?
1.3.2.10-Are there any problems for lot applicants?

1.3.3. Housing production

1.3.3.1- Has there ever been a social housing project in Daloa?
1.3.3.2- Do lot owners build quickly?
1.3.3.3- When a lot is not highlighted, what do you do?
1.3.3.4- How many lots have you already withdrawn?
1.3.3.5- What is the deadline for withdrawing an awarded lot?
1.3.3.6- Do you impose a type of housing? A type of building material?
Which ones?

1.3.3.7- If the instructions are not followed, what do you do?
1.4. INTERVIEW WITH THE REGIONAL DIRECTOR OF CONSTRUCTION AND

OF URBANISM

Date and place of interview

1.4.1.   What is your job (role of the construction department in land management)?

1.4.2.   How is   the relief   of   the
city?
1.4.3.   Do you have the master plan of the city of Daloa?

1.4.4.   What types of houses are imposed in Daloa?

1.4.5.   Do you have the possibility to check the house plans on the spot?

1.4.6.   Do you have the right staff to carry out your mission?

1.4.7.   What is a technical file made of?

1.4.8.   What types of house plans do you allow?

1.4.9.   Are your expectations met on the ground?

1.4.10.   Do you follow exactly the legislator's policy on housing?

1.4.11.   What is the actual area developed in Daloa? Before 2104/After
2014 /   Total
1.4.12.   Have you ever sat on a committee for the withdrawal of a lot?

1.4.13.   What is the estimated number of lots withdrawn?
1.4.14.   How many lots have been regularly awarded so far?
1.4.15.   How do you become a landowner in Daloa?
1.4.16.   What do you think of the proliferation of informal settlements and precarious housing in the commune
of Daloa?
1.4.17.   What are your difficulties?
1.4.18.   Identification of purchasers Sex ... Age ...   Ethnicity/NationalityOccupationLevel of   education

1.4.19. What are the areas of land disputes in Daloa?
1.4.20. What are the land conflicts in Zakoua?

1.4.21. What are the land conflicts in Tagoura and Sapia?

1.4.22. What is the method of conflict resolution?
1.4.23. How are conflicts resolved?

1.4.24. What is the status of collaboration with village communities?

1.4.25. What is the place of indigenous people before and after 2014?

1.4.26. Does the management take into account the customs of indigenous people in land management?
    YesNo
If so, how?
1.4.27. How does the CDA reform in Daloa work?

1.4.28. What are the difficulties encountered by the construction department in land management?
1.4.29. What is the procedure for acquiring urban land in Daloa?

1.4.30. Which technical services are involved in land management in Daloa?

1.4.31. Who are the actors in the developments?
1.4.32. What type of housing estate is practical? a- village b- administrative c- private
1.4.33. What is the subdivision procedure?

Table 18: Distribution of net population density in Daloa in 2014

Order number	Neighbourhoods	Population	Area (ha)	Density^ (inhab/ha)
1	Slaughterhouse 1	14 559	66	221
2	Aviation	6 234	47	133
3	Baouɪë	4 526	30	151
4	Belleville	11564	47	246
5	Trade	4 560	93	49
6	Dalo-labia	7 822	68	115
7	Dioulabougou	14 333	77	186
8	E\'ëc11ë	704	120	6
9	Gbeuliville	5 461	30	182
10	Gbobëlë	13 285	59	225
11	Huberson	13 157	80	164
12	Kennedy	7 464	117	64
13	Kirman	1 433	73	20
14	Labia	4 557	71	64
15	Lobia 1	3 205	52	62
16	Lobia 2	8 292	90	92
17	Marshes	15 292	51	300
18	Mossidougou	9 269	30	309
19	Orly	8 458	111	76
20	Orly Military Camp	18 850	108	174
21	Orly Squadron	5 666	138	41
22	Wolof	2 064	17	121
23	Swimming pool	2 376	52	46
24	Segou	1 499	30	50
25	Sun	9 010	52	173
26	Sun 2	6 567	99	66
27	South A (Garage Zai)	12 460	67	186
28	South B (Slaughterhouse 2)	14 723	81	182
29	South C (Lycëe Fadiga)	1 292	72	18
30	Sud D (Soap factory)	5 250	85	62
31	Tazibouo	7 755	96	81
32	Tazibouo French School	976	40	24
33	Tazibouo General Staff	2700	52	52

Source: RGPH 2014

6.1. The environmental impact of urban dynamics on management

Buy your books fast and straightforward online - at one of world's fastest growing online book stores! Environmentally sound due to Print-on-Demand technologies.

Buy your books online at
www.morebooks.shop

Kaufen Sie Ihre Bücher schnell und unkompliziert online – auf einer der am schnellsten wachsenden Buchhandelsplattformen weltweit! Dank Print-On-Demand umwelt- und ressourcenschonend produzi ert.

Bücher schneller online kaufen
www.morebooks.shop

Printed by Books on Demand GmbH, Norderstedt / Germany